Springer Proceedings in Materials

Springer Proceedings in Materials publishes the latest research in Materials Science and Engineering presented at high standard academic conferences and scientific meetings. It provides a platform for researchers, professionals and students to present their scientific findings and stay up-to-date with the development in Materials Science and Engineering. The scope is multidisciplinary and ranges from fundamental to applied research, including, but not limited to:

- Structural Materials
- Metallic Materials
- Magnetic, Optical and Electronic Materials
- Ceramics, Glass, Composites, Natural Materials
- Biomaterials
- Nanotechnology
- Characterization and Evaluation of Materials
- Energy Materials
- Materials Processing

To submit a proposal or request further information, please contact one of our Springer Publishing Editors:

China: **Mengchu Huang** (mengchu.huang@springer.com)
India: **Akash Chakraborty** (akash.chakraborty@springernature.com)
Europe: **Mayra Castro** (mayra.castro@springer.com)
America: **Michael Luby** (Michael.luby@springer.com)

More information about this series at http://www.springer.com/series/16157

Francisco José Galindo-Rosales ·
Laura Campo-Deaño · Alexandre M. Afonso ·
Manuel A. Alves · Fernando T. Pinho
Editors

Proceedings of the Iberian Meeting on Rheology (IBEREO 2019)

Springer

Editors
Francisco José Galindo-Rosales
Department of Chemical Engineering
Faculty of Engineering
University of Porto
Porto, Portugal

Laura Campo-Deaño
Department of Mechanical Engineering
Faculty of Engineering
University of Porto
Porto, Portugal

Alexandre M. Afonso
Department of Mechanical Engineering
Faculty of Engineering
University of Porto
Porto, Portugal

Manuel A. Alves
Department of Chemical Engineering
Faculty of Engineering
University of Porto
Porto, Portugal

Fernando T. Pinho
Department of Mechanical Engineering
Faculty of Engineering
University of Porto
Porto, Portugal

ISSN 2662-3161 ISSN 2662-317X (electronic)
Springer Proceedings in Materials
ISBN 978-3-030-27703-1 ISBN 978-3-030-27701-7 (eBook)
https://doi.org/10.1007/978-3-030-27701-7

This Springer imprint is published by the registered company Springer Nature Switzerland AG
The registered company address is: Gewerbestrasse 11, 6330 Cham, Switzerland

Preface

The 7th edition of the Iberian Meeting on Rheology (IBEREO 2019) was held in Porto (Portugal) on September 4–6, 2019. The meeting is an international event organized every two years, jointly by the Spanish Group of Rheology (GER) and the Portuguese Society of Rheology (SPR). The current edition took place in the Rectory Building of the University of Porto, located at the historical center of the city of Porto.

During the meeting, recent trends in rheology were addressed, with a special emphasis on both basic science and industrial applications covered from different perspectives, i.e., experimental, theoretical, and numerical. The meeting included two keynote lectures, as well as oral and poster contributions.

This manuscript gathers a selection of the contributions to the conference, which intends to illustrate the state-of-the-art knowledge and experience of a multidisciplinary group of scientists that have in common their research activities on the topic of rheology, using theoretical analysis, experimental characterization, and numerical simulation approaches. The primary audience of this work are the researchers working in academia or in industry, but this text is also of relevance, as supplementary reading, for anyone working with complex fluids, fluid-flow characterization techniques and numerical simulations.

The contents of this book are organized in several topics that coincided with the following Sessions of the IBEREO 2019 meeting:

- Food, Cosmetics and Pharmaceutical Products
- Polymers and Biopolymers
- Rheometry and Experimental Methods
- Non-Newtonian Fluids Mechanics
- Suspensions and Colloids
- Microfluidics and Microrheology

September 2019

The Organizing Committee
Iberian Meeting on Rheology 2019

Organization

Organizing Committee

The 7th edition of the Iberian Meeting on Rheology was locally organized by the following members of the Transport Phenomena Research Center, a Research Unit of the Faculty of Engineering of the University of Porto, focused on the sub-domain of transport phenomena (heat, mass, and momentum) bridging the areas of chemical and mechanical engineering:

Francisco J. Galindo-Rosales
Department of Chemical Engineering, Faculty of Engineering,
University of Porto, Rua Dr. Roberto Frias s/n, 4200-465 Porto, Portugal,
e-mail: galindo@fe.up.pt

Laura Campo-Deaño
Department of Mechanical Engineering, Faculty of Engineering,
University of Porto, Rua Dr. Roberto Frias s/n, 4200-465 Porto, Portugal,
e-mail: campo@fe.up.pt

Alexandre M. Afonso
Department of Mechanical Engineering, Faculty of Engineering,
University of Porto, Rua Dr. Roberto Frias s/n, 4200-465 Porto, Portugal,
e-mail: aafonso@fe.up.pt

Manuel A. Alves
Department of Chemical Engineering, Faculty of Engineering,
University of Porto, Rua Dr. Roberto Frias s/n, 4200-465 Porto, Portugal,
e-mail: mmalves@fe.up.pt

Fernando T. Pinho
Department of Mechanical Engineering, Faculty of Engineering,
University of Porto, Rua Dr. Roberto Frias s/n, 4200-465 Porto, Portugal,
e-mail: fpinho@fe.up.pt

Scientific Committee

Each contribution to this book of proceedings was reviewed by two of the members of the Scientific Committee of the Iberian Meeting on Rheology 2019:

Maria Teresa Cidade (SPR President)	Universidade Nova de Lisboa, Portugal
Antonio Guerrero (GER President)	Universidad de Sevilla, Spain
Isabel de Sousa	Universidade de Lisboa, Portugal
Francisco J. Rubio-Hernández	Universidad de Málaga, Spain
Catarina Rosa Leal	Instituto Politécnico de Lisboa-ISEL, Portugal
José María Franco Gómez	Universidad de Huelva, Spain
João Miguel Nóbrega	Universidade do Minho, Portugal
Antxón Santamaría	Universidad del País Vasco, Spain
José Covas	Universidade do Minho, Portugal
María Jesús Hernández Lucas	Universitat de València, Spain
Maria Graça Rasteiro	Universidade de Coimbra, CIEPQF-UC, Portugal
Clara A. Tovar	Universidade de Vigo, Spain
Paulo J. Oliveira	Universidade da Beira Interior, Portugal
João Maia	Case Western Reserve University, USA
José Maria Ferreira	Universidade de Aveiro, Portugal

Contents

Polymers and Biopolymers

Rheometry and Experimental Methods

Food, Cosmetics and Pharmaceutical Products

Soy Protein-Based Matrices with Incorporated Salts for Use in Horticulture

Antonio Guerrero[1(✉)], Mercedes Jiménez-Rosado[1], Víctor Pérez-Puyana[2], Felipe Cordobés[1], and Alberto Romero[2]

[1] Technology and Design of Multicomponent Products, Escuela Politécnica Superior, University of Seville, C/Virgen de África, 7, 41011 Seville, Spain
{aguerrero,mjimenez42,fcordobe}@us.es

[2] Technology and Design of Multicomponent Products, Facultad de Química, University of Seville, C/Profesor García González s/n, 41012 Seville, Spain
{vperezll,alromero}@us.es

Abstract. Nowadays, the use of fertilizers has become widespread in horticulture, causing pollution problems in soils and groundwater due to the low assimilation by the plants and their high solubility. The incorporation of these fertilizers in biodegradable matrices, which can release them in a controlled manner, could solve all these disadvantages. Furthermore, these matrices can cause other advantages, such as water retention that can be used in drought periods. The objective of this work was the evaluation of the incorporation of different zinc (micronutrient) salts into soy protein-based matrices. Thus, the water uptake capacity, degree of zinc incorporation and mechanical properties of these matrices were compared.

1 Introduction

Horticultural overproduction has generated a high degradation of the farmland, by the depletion of soil nutrients, which makes it necessary to use fertilizers to maintain the production [1]. However, an excessive use of fertilizers can cause subsoil and groundwater contamination due to their high solubility [2]. A potential solution to this problem is the incorporation of nutrients in biodegradable matrices, which can be available for convenient controlled-release depending on the needs of the crops. The use of protein-based matrices, such as soy bean protein, can also incorporate nitrogen (from the high protein content), retain water and supply it in drought periods [3].

In this way, the main objective of this work was the incorporation of different zinc salts (zinc sulfate heptahydrate ($ZnSO_4{\cdot}7H_2O$), zinc sulfate monohydrate ($ZnSO_4{\cdot}H_2O$), zinc chelated with EDTA [2, 2′, 2″, 2‴-(Ethylene-1, 2-diildinitrile)tetraacetic acid] (Zn EDTA) and zinc perchlorate ($Zn(ClO_3)_2$) to soy protein-based matrices. The water uptake capacity, degree of zinc incorporation and mechanical properties of these matrices were compared. The results showed the excellent potential of soy protein-based matrices for the incorporation of micronutrients (zinc). On the other hand, the incorporation of salts into the matrices caused them to decrease their water uptake capacity,

F. J. Galindo-Rosales et al. (Eds.): IBEREO 2019, SPM, pp. 3–6, 2020.
https://doi.org/10.1007/978-3-030-27701-7_1

although, according to their rheological analysis, all the obtained matrices had suitable characteristics for their subsequent scaling and industrialization.

2 Materials and Methods

2.1 Materials

Soy protein isolate (SPI, min. 91% protein) was used as the biopolymer, supplied by Protein Technologies International (SUPRO 500E, Belgium). The plasticizer was glycerol (Gly) provided by Panreac Química S.A. (Spain). The different salts used were $ZnSO_4 \cdot 7H_2O$, $ZnSO_4 \cdot H_2O$ and $Zn(ClO_4)_2$ supplied by Panreac Química S.A. and Zn EDTA, provided by Trade Corporation International S.A.U. (Spain).

2.2 Matrices Fabrication Technique

The matrices were prepared by a generic matrix preparation [3]. SPI and Gly in a 1:1 ratio with the different salts were first homogenized in a rotating mixer Polylab QC (ThermoHaake, Germany) at 50 rpm for 10 min. The salt incorporated was the maximum percentage of salt with which the blends could be processed. Later, the dough-like blends were processed by injection molding using a MiniJet Piston Injection Molding System II (ThermoHaake, Germany), following the same protocol as in previous studies [3]. Finally, the bioplastic matrices were subjected to a dehydrothermal treatment in an oven at 50 °C for 24 h.

2.3 Matrices Characterization

Water Uptake Capacity

Water uptake capacity tests were performed according to the ASTM D570 standard [4], using rectangular bioplastic matrices (55 × 10 × 1 mm).

Inductively Coupled Plasma-Atomic Emission Spectroscopy (ICP-AES)

Water uptake can cause part of the salt to be removed. For this reason, it is necessary to estimate the amount of micronutrient (Zn) that will be retained in the matrix or be removed. To this end, ICP-AES measurements were carried out in an ICP SpectroBlue TI (Spectro, Germany).

Mechanical Properties

The matrices with incorporated micronutrients must have some mechanical resistance for their subsequent scaling to the industrial level. For this reason, the mechanical properties of the matrices were measured with compression tests. Thus, frequency sweep tests were performed from 0.02 to 20 Hz at room temperature and a constant strain of 0.01% (within the viscoelastic range) obtaining the compressional elastic (E′) and loss (E″) moduli.

3 Results and Discussion

Table 1 collects the data obtained for the water uptake capacity tests and ICP-AES Zn^{2+} measurements. The water uptake capacity tests show that the incorporation of salt into the matrices led to the loss of their superabsorbent capacity, except for Zn EDTA, which not only maintained its capacity but improved it. This behavior can be due to the ionic interactions created with the dissolution of the salts in the water. When Zn EDTA was incorporated, these ionic interactions were lower because the EDTA complex retains Zn in the medium.

According to the Zn^{2+} retained, the matrix with 10% $ZnSO_4{\cdot}H_2O$ was the one that held the highest Zn^{2+} percentage, both before and after water absorption. On the other hand, a higher addition of salt led to a greater loss of zinc. However, the matrix with 10 wt% Zn EDTA was the one that had the highest zinc loss. This loss could be due to a low capability of the matrix to retain the zinc, which is thereby released together with EDTA. In the other matrices, when ionizing, the positive charges of zinc ions interact with the negative charges of the protein, thus holding them in the matrix.

Table 1. Water uptake capacity (WUpC) and ICP-AES Zn^{2+} measurements for the different bioplastic matrices. The different letters represent significant differences

Systems	WUpC (%)	Zn^{2+} in matrices (wt%)	
		Before WUpC	After WUpC
Reference	950^{a}	-	-
2.5% $ZnSO_4{\cdot}7H_2O$	387^{b}	0.56^{a}	0.46^{e}
10% $ZnSO_4{\cdot}H_2O$	98^{c}	3.62^{b}	1.84^{f}
10% Zn EDTA	1356^{a}	1.62^{c}	0.09^{g}
10% $Zn(ClO_4)_2$	111^{c}	2.46^{d}	1.82^{h}

The mechanical properties of the different matrices are shown in Fig. 1. It can be observed that the incorporation of salt into the soy protein-based matrices reduced the elastic modulus. In this way, the matrices were more rigid when salt was incorporated. However, the loss tangent did not undergo significant changes. Thus, the relationship between the elastic and viscous moduli was maintained regardless of the amount of incorporated salt.

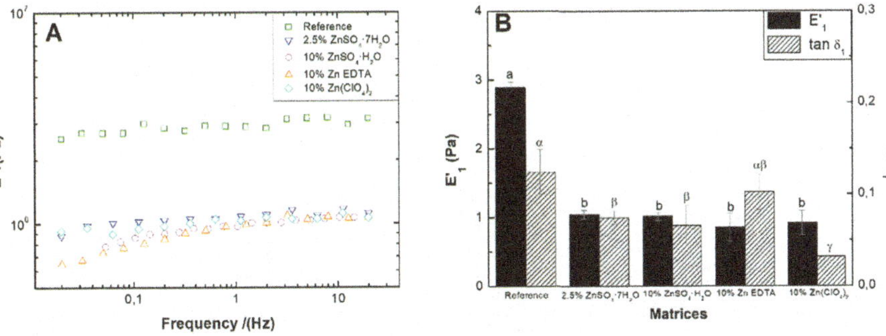

Fig. 1. A: Elastic modulus (E′) vs. Frequency, B: Elastic modulus (E'_1) and loss tangent (tan δ_1) values at 1 Hz. The different letters represent significant differences.

4 Conclusions

To sum up, soy protein-based matrices have demonstrated their potential as a vehicle for the incorporation of micronutrients to plants in a controlled manner. Furthermore, these matrices can retain water and supply it in drought periods and have the necessary mechanical properties for their subsequent scaling to the industrial level.

Acknowledgements. The authors acknowledge the financial support from MINECO/FEDER, EU for the research project (Ref. CTQ2015-71164-P). In addition, authors also appreciate the pre-doctoral fellowships of V. Pérez Puyana (VPPI-US) and M. Jiménez Rosado (FPU17/01718).

References

1. Hazell, J.L., Wood, S.: Drivers of change in global agriculture. Philos. Trans. R. Soc. B Biol. Sci. **236**, 495–515 (2008)
2. Muscanescu, A.: Organic versus conventional: advantages and disadvantages of organic farming. Econ. Eng. Agric. Rural Dev. **13**, 253–256 (2013)
3. Jiménez-Rosado, M., Pérez-Puyana, V., Cordobés, F., Romero, A., Guerrero, A.: Development of soy protein-based matrices containing zinc as micronutrient for horticulture. Ind. Crops Prod. **121**, 345–351 (2018)

Emollient and Surfactant Influence in Microemulsions Rheology

Andreia Nunes[1(✉)], Joana Marto[2], Francisca Lopes[2], and Helena Margarida Ribeiro[2]

[1] Faculty of Pharmacy, Universidade de Lisboa, Lisbon, Portugal
andreia.a.nunes@campus.ul.pt
[2] Research Institute for Medicines (iMed.ULisboa), Faculty of Pharmacy, Universidade de Lisboa, Lisbon, Portugal
jmmarto@ff.ulisboa.pt, fclopes@ff.ul.pt, hribeiro@campus.ul.pt

Abstract. Microemulsions are selected for topical delivery because of their stability and isotropic nature. Although microemulsion is macroscopically homogeneous, microscopically, it can be seen the presence of different structures such as oil droplets in water, water droplets in oil, bicontinuous, lamellar mixtures, etc., which can affect the rheology of these systems. The aim of this work was to study the influence of the amount of emollient and surfactant in the rheology of microemulsions for topical delivery.

1 Introduction

Skin is used as a route to deliver active agents (topical delivery), allowing the treatment of skin diseases. Thus, microemulsions-based drug delivery systems are a very interesting alternative not only to overcome poorly water-soluble drugs, but also to adjust the drug delivery kinetics [1, 2]. Microemulsions (ME) can be defined as isotropic and thermodynamically stable oil-in-water (O/W) or water-in-oil (W/O) emulsions. These formulations are mainly constituted by water, surfactants and oils. However, this system presents some disadvantages when designed for topical use, such as, the low viscosity, resulting in the tendency to drip from the skin surface, not having an easy application. Thus, ME are frequently transformed into complex semi-solid systems. Depending on the percentages used of each component, microemulsions may present different macroscopic aspects, being liquid or viscous [2]. These systems allow the solubilization of hydrophilic and lipophilic actives with no input energy.

According to Winsor [3] in 1948, four types of microemulsions can be defined. The first type (i) can be characterized by a two phase system (biphasic), where the upper phase presents an excess of oil and the lower phase presents an O/W emulsion. Second type (ii) also has a two-phase system, where the less dense phase contains W/O emulsion and the denser phase contains excess of water. Third type (iii) presents a single microemulsion phase (monophasic) and, finally, the fourth type (iv) exhibit a three-phase system where the upper phase contains an excess of oil, the middle phase presents a bicontinuous microemulsion and the lower phase shows an excess of water [4].

F. J. Galindo-Rosales et al. (Eds.): IBEREO 2019, SPM, pp. 7–11, 2020.
https://doi.org/10.1007/978-3-030-27701-7_2

The aim of this work was to study the influence of the amount of emollient and surfactant in the rheology of microemulsions for topical delivery.

2 Materials and Methods

Two microemulsions were prepared: ME 1 with 40.30% of Cithrol™ 10GTIS (PEG-20 Glyceryl triisostearate), 47.60% of Emosmart™ C28 (C21-28 Alkane), 5.97% Propylene Glycol EP (Propylene Glycol) and 5.97% of water. The ME 2 containing 50.00% of Cithrol™ 10GTIS, 5.00% of Emosmart™C28, 5.00% Propylene Glycol EP and 39.90% of water. These formulations were further characterized regarding pH and rheology.

2.1 pH

The pH values were determined using a potentiometer (METTLER TOLEDO) after 24 h at a room temperature (25 °C).

2.2 Rheology

The rheological characterization was performed with viscosity and oscillatory experiments with a controlled stress Kinexus Rheometer (Malvern) using a cone-plate geometry (cone angle 4° and radius 40 mm). All the studies were performed at room temperature (25 °C). In the continuous shear experiment, the shear stress of each formulation was obtained by increasing the shear rate from 0.1 s^{-1} to 100 s^{-1} and 8 points per decade during 100 s. In the oscillatory method, an amplitude sweep test where the shear strain was between 0.01 Hz and 100 Hz, the frequency was 1 Hz and 10 points per decade. Then, a frequency sweep test was performed with a shear strain of 0.05% for the viscous microemulsion and 5.0% for liquid microemulsion, and a frequency between 0.01 Hz and 10 Hz.

3 Results and Discussion

The formulations obtained in this study were liquid (ME 1) and viscous (ME 2).

3.1 pH

The pH values vary from 7.0 to 7.7. The pH values obtained are according to the pH of each of the components used.

3.2 Rheology

The two formulations were analyzed by observing the behavior of viscosity as a function of shear rate applied and the results are present in Fig. 1. The results show that the viscous formulation (ME 2) presents a shear thinning behavior, while the liquid formulation (ME 1) has lower viscosity values and shows a Newtonian flow.

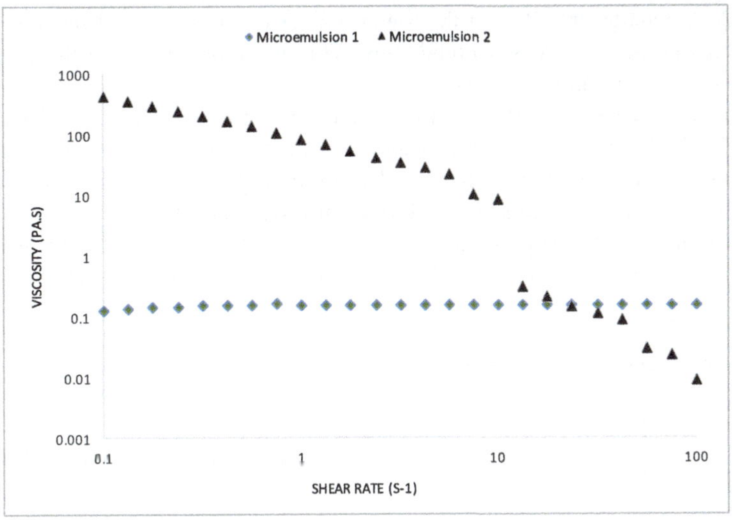

Fig. 1. Microemulsions flow curves

The construction of phase diagrams, and according to Winsor [3], ME 1 and ME 2 are W/O surfactant-rich oil phase. Nevertheless, in ME 2, other structures, such as lamellar or bi continuous phases are probably present, because in this formulation there is higher content in surfactant (Fig. 2).

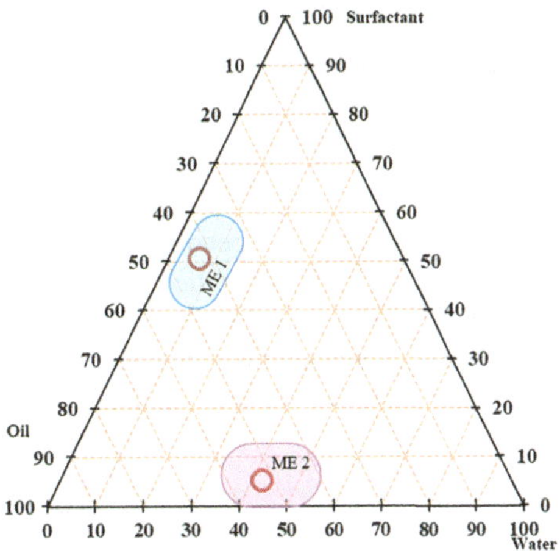

Fig. 2. Ternary diagram of microemulsions

The oscillatory method allows to analyze the behavior of elastic and viscous component of each ME as a function of frequency (Fig. 3). The results showed that the ME 1 has the loss modulus higher than the storage modulus meaning a liquid like structure, while ME 2 showed the storage modulus higher than the loss modulus thus, has solid-like character. ME 2 at the lowest stresses shows a solid behavior, while in higher stresses demonstrates a liquid flow. This is probably due to the presence of lamellar and bi-continuous phases.

With the increase of the stress applied there is a disruption of the micellar structures present in this formulation. Nevertheless, ME 2 permits a good spreadability and adhesion of this formulation on the skin. Regarding to loss tangent the values obtained are 10.95 and 0.42 for a frequency of 0.01 Hz for ME 1 and ME 2, respectively, which are in accordance with the previous results. The surfactant and emollient concentration affect the structures and consequently the rheology of the formulations developed.

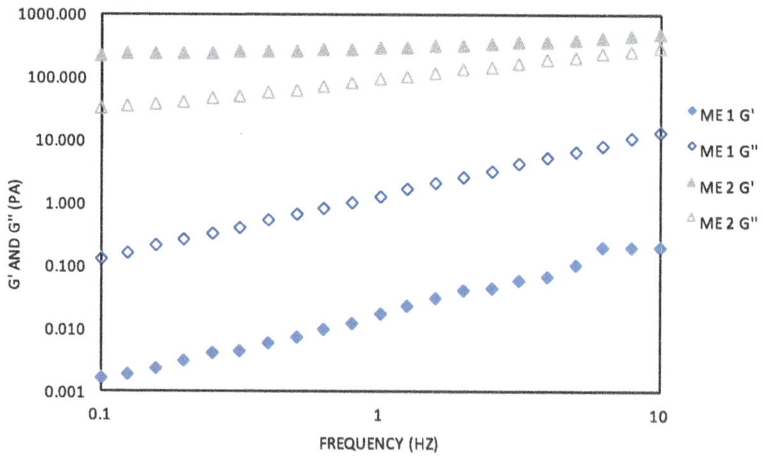

Fig. 3. Microemulsions oscillatory results

4 Conclusions

The results so far obtained demonstrate that depending of the amount of each ingredient (emollient and surfactant) the rheological properties of these microemulsions are different, affecting their adhesion and spreadability. Although, liquid microemulsions present less adhesion and higher spreadability, they also present a lower amount of surfactant in the formulation, which can be safer to the skin.

References

1. Lopes, L.B.: Overcoming the cutaneous barrier with microemulsions. Pharmaceutics **6**, 52–77 (2014)
2. Froelich, A., Osmałek, T., Snela, A., et al.: Novel microemulsion-based gels for topical delivery of indomethacin: formulation, physicochemical properties and in vitro drug release studies. J. Colloid Interface Sci. **507**, 323–336 (2017)
3. Winsor, P.A.: Hydrotropy, solubilisation and related emulsification processes. Trans. Faraday Soc. **44**, 376–398 (1948)
4. Callender, S.P., Mathews, J.A., Kobernyk, K., Wettig, S.D.: Microemulsion utility in pharmaceuticals: implications for multi-drug delivery. Int. J. Pharm. **526**, 425–442 (2017)

Fat Reduction in Extruded Snacks with the Use of Xanthan Gun

Diego R. Marques[1], Kimberli P. Berwig[1], Carla Graça[2], Antonio R. G. Monteiro[1(✉)], and Isabel Sousa[2]

[1] Department of Food Engineering, Universidade Estadual de Maringa, Maringá, Brazil
diegomarques86@gmail.com, kim.berwig@gmail.com, argmonteiro@uem.br

[2] LEAF – Linking Landscape, Environment, Agriculture and Food, Instituto Superior de Agronomia, Universidade de Lisboa, Lisbon, Portugal
carlalopes.graca86@gmail.com, isabelsousa@isa.ulisboa.pt

Abstract. Food industries are modifying their products and processes in order to adapt to the needs and desires of consumers for a healthy diet. Extruded snacks are products where 10–20% of fat is sprinkled on the product to fix flavors, seasonings and salts. Considering the need for flavoring snacks and simultaneously reducing of the calories intake, an alternative to fat, based on polysaccharides, was studied.

This work aimed to evaluate the use of aqueous gum solutions to replace the oil in the flavoring of snacks, based on texture profile of the snacks and rheology features of the coatings. Texture profile was applied in compression mode to evaluate the hardness of the coated snacks. Flow behaviour of coating solutions at different pH was determined. The moisture, water activity and refraction index of the coated snacks, was also evaluated.

Results for the aqueous coatings were very encouraging showing good coating properties and not damaging texture of the extrudates, comparing well with the oil based coatings for sensory evaluation.

1 Introduction

The extrusion process is used to produce several foods such as animal breakfast cereal-based food, pasta and extruded snack. Cereal flours are the main raw material for these foody applications.

The manufacture of extruded salty snacks consists of extruding pieces of degeminated corn (corn grits) after which, flavouring is performed. The aromatization aims to make the product more attractive sensorily, by sprinkling of 10 to 20% (w/w) vegetable fat or vegetable oil on the extrudates followed by the addition of aromas, seasonings and salt [1].

Xanthan gum (Xg) is a polysaccharide of microbial origin produced by species of bacteria of the genus *Xanthomonas*. It is soluble in hot or cold water. It has stable viscosity in solutions with pH ranging from 2–11 and also at temperatures up to 90 °C.

F. J. Galindo-Rosales et al. (Eds.): IBEREO 2019, SPM, pp. 12–15, 2020.
https://doi.org/10.1007/978-3-030-27701-7_3

The Xanthan gum forms a viscous solution applying low shear forces and has a pseudoplastic characteristic that is, viscosity decreases with increasing deformation rate but can recover its viscosity by suspending shear [2].

This work intends to evaluate the use of aqueous gum solutions to replace the oil in the flavoring of snacks, based on texture profile of the snacks and rheology features of the coatings.

2 Materials and Methods

The snacks used were prepared under industrial conditions using only maize grits (ProVida, Portugal). For the coating of the snacks, soybean oil was used as a control (SONAE, Portugal) and an aqueous solution of Xanthan gum (SOSA, Spain) at pH 3.5 and 7.0; the pH was adjusted by citric acid (Merck, Germany). The Xg aqueous solution was kept under agitation for 10 min and then storage in room temperature for 7 h before use.

2.1 Coating

The flavoring was performed by coating, pouring the solutions over the snacks in agitation followed by adding a mixture of dehydrated herbs and salt (25% Parsley, 15% oregano, 10% basil and 50% sodium chloride). Seven treatments using Xg at different levels (0.25, 0.5 and 1.0%) and two pH values (3.5 and 7.0), were performed. Soybean oil solution (10%) was used as the control. Condition for flavor coating: 100 g of the snacks were placed in the drawer, then 10 ml of the sprinkling was sprayed and left to homogenize for 30 s, then 4 g of the seasoning was again homogenized for 30 s, then the snacks were dried for 35 min, at 65 °C and cooled to room temperature.

2.2 Texture Analysis

Texture evaluation of the snacks (hardness) was performed using the texturometer TA-XTplus (Stable MicroSystems, UK), in compression mode. An acrylic cylindrical probe with 25 mm diameter (P/25L) at 1 mm/s of test-speed, with a load cell of 5 kg, was used.

2.3 Rheology

The coating materials were submitted to a rheological characterization using a controlled-stress rheometer (Haake Mars III – Thermo Scientific, Germany) coupled to a UTC-Peltier system for temperature control. The flow behaviour at steady shear was evaluated using a 2°cone and plate sensor system with 35 mm diameter was used. All assays were performed at 20 °C, and were repeated at least three times.

2.4 Moisture, Water Activity and Retraction Index

Moisture was determined by oven drying at 105 °C (AACC 44-15.02). The Hygrolab (Rotronic, UK) was used to determine the water activity of the snacks, at constant temperature (20.0 ± 0.2 °C). The retraction index was determined by the weight ratio: coated snacks/uncoated snacks weight as described by Monteiro *et al.* [1].

2.5 Sensorial Analysis

Sensory evaluation was performed by 105 untrained tasters using hedonic scale with 9 point, from 1 unlike very much until 9 like very much.

3 Results and Discussion

The effect of the flavoring coating solution (pH 3.5 and 7.0) was evaluated on snacks coated hardness.

From Fig. 1 it can be seen that the hardness ranged from 44.7 to 53.5 N, and significant differences were only observed at snacks coated with coating solutions at pH 7.0 (Fig. 1).

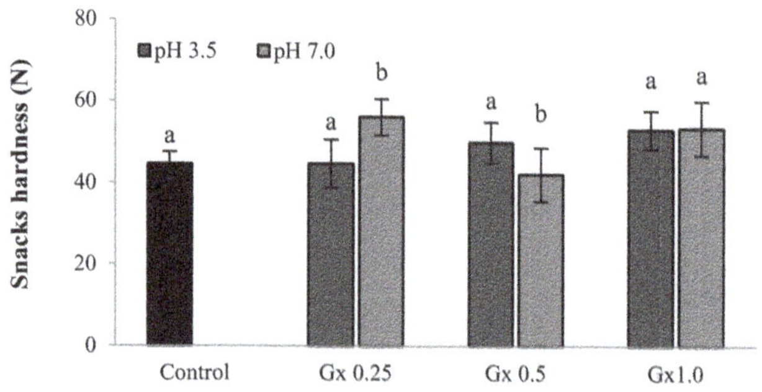

Fig. 1. Evaluation of the snacks hardness with coating solutions at pH 3.5 and 7.0

Flow measurements, to evaluate the effect of the different percentage of the Xg and the impact of the pH values applied on flow behaviour, was performed. As it can be seen from Fig. 2, all samples presented a typical shear-thinning behavior and the different pH values applied have no impact on viscosity. From these results one can say that pH 3.5 is most favorable to replace oil by xanthan aqueous coatings being this pH also better in terms of preservation of the products.

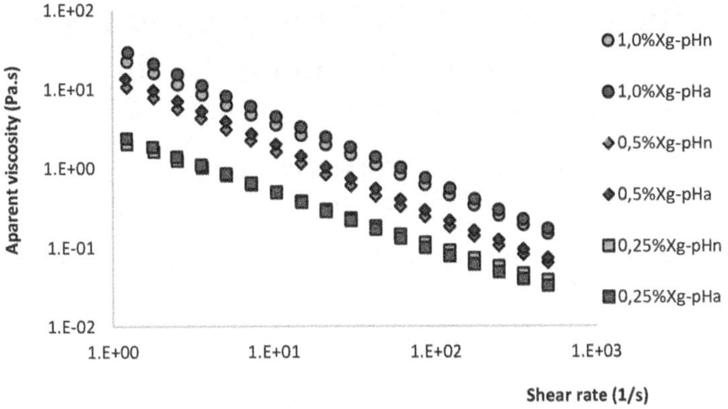

Fig. 2. Viscosity *Vs* shear-rate curves, at 20 °C for samples with 1.0, 0.5 and 0.25% of xanthan Gum (Xg) at different pH values: 3.5 (pHa) and 7.0 (pHn)

In terms of physic-chemical properties, the moisture ranged from 4.97–5.18% and water activity from 0.187 to 0.198. The retraction index was higher than 0.98 in all samples. These results show that the water present is too low to interfere with the preservation of the final product and is a typical characteristic of these crunchy starch texturized products.

In the sensory analysis, a high average acceptance was obtained (from 6.8 to 7.6), not having significant differences between all of xanthan gum samples and the control, meaning that there were no differences in terms of acceptance in all samples.

All results did not present significant differences, even when in comparison with the standard for all characteristics: texture, flow behaviour, moisture, water activity and retraction index, indicating that the replacement of soybean oil by aqueous solution of xanthan gum is highly feasible.

4 Conclusions

The flavored snacks, using xanthan gum solutions presented similar characteristics in terms of texture, flow behaviour, moisture and a_w compared to the snacks flavored with fat, concluding that it is possible to replace the lipids in the aromatization of snacks obtaining an excellent final product, with fat and calorie reduction and subsequent health benefits.

References

1. Monteiro, A.R.G., Marques, D.R., Marchi, L.B., Chinellato, M.M., Berwig, K.P., Wolf, B.: Eliminating the use of fat in the production of extruded snacks by applying starch coating. Chem. Eng. Trans. **49**, 625–630 (2016)
2. Rosalam, S., England, R.: Review of xanthan gum production from unmodified starches by Xanthomonas campestris sp. Enzyme Microb. Technol. **39**(2), 197–207 (2006)

Interfacial and Bulk Rheology of Food Emulsions Containing Inulin

Carlos Bengoechea(✉), María Luisa López-Castejón, Manuela Ruiz-Domínguez, and Cecilio Carrera

Chemical Engineering, Polytechnic School, University of Seville, C/Virgen de África, 7, 41011 Seville, Spain
{cbengoechea, llcastejon, manuela, cecilio}@us.es

Abstract. Inulin is a natural polysaccharide that is commonly used as food ingredient due to its noteworthy functional and prebiotic properties, as well as biological effects. Moreover, inulin may be used as a thickening agent improving the kinetical stability of thermodynamically unstable food emulsions through an increase in the viscosity of the continuous phase, their yield stress and inhibiting the droplets mobility. On these terms, the present work deals with an interfacial and bulk characterization of the rheology of Oil in Water emulsions (O/W: 5/95) stabilized by β-lactoglobulin and inulin. Emulsions stabilized by β-lactoglobulin (0.5% wt.), with different inulin contents (0 and 10% wt.) were prepared using a high-pressure valve homogenizer. Physical-chemical properties are influenced by both the interactions that may take place among inulin and β-lactoglobulin molecules and by the thickening effect of inulin present in the continuous phase. Depending on the inulin content either the former or the latter may prevail. At 10% wt. inulin content, the inulin thickening character was more important than the potential effects of thermodynamical incompatibility among the protein and the polysaccharide.

Therefore, the importance of the interactions of inulin with the emulsifier used, as well as its role as thickener, prove that they are key factors to consider when including inulin in commercial food emulsions. Thus, a 10% wt. inulin content leads to stable emulsions even if the interfacial film possesses poorer viscoelastic properties.

1 Introduction

There is an increasing interest to reduce or eliminate fat in foods, especially through the use of fat replacers with thickening and gelling properties. For this purpose, it is common the use of carbohydrates with additional healthy properties like inulin. These ingredients act as stabilizers enhancing the emulsion stability by increasing the viscosity of the continuous phase, their yield stress and inhibiting the drop-lets mobility [1]. Moreover, protein-polysaccharide interactions can play a significant role in the structure and stability of many processed foods, sometimes forming hard and permanent 'conjugates' with proteins. Functional properties of food proteins, such as surface activity, solubility or emulsifying and foaming properties are also affected by their interactions with polysaccharides, determining the structure-property relationship in

F. J. Galindo-Rosales et al. (Eds.): IBEREO 2019, SPM, pp. 16–19, 2020.
https://doi.org/10.1007/978-3-030-27701-7_4

food systems. In this sense, the control of these interactions is very important for the development of novel food processed products. Inulin is a polysaccharide recognized as a safe food ingredient with interesting biological effects, being a potent complement pathway activator when in a particulate form and having anti-cancer properties [2]. In fact, due to its chemical structure, inulin is neither hydro-lysed nor adsorbed in the small intestine but is extensively fermented by the colon bacteria [3].

Rheological properties, both bulk and interfacial, are very important for the development of new products, estimation of its shelf-life, sensory assessment, and stability evaluation of the food product. These properties can be affected by some factors like temperature, concentration of continuous phase, and pH.

The main objective of this work is to study the effect of inulin on interfacial and flow properties of O/W emulsions stabilized by a protein emulsifier (β-lactoglobulin). For this, the influence of inulin concentration in β-lactoglobulin-inulin solutions is studied analysing adsorption kinetics of the interfacial films adsorbed at the O/W interface and rheological characteristics of the bulk emulsions, as well as the short-term physical stability of emulsions (up to 15 days).

2 Materials and Methods

2.1 Manufacture and Sampling

In the present work, inulin (degree of polymerization: 36; free glucose and fructose ($\leq 0.05\%$)) and β-lactoglobulin ($\geq 90\%$) were provided from Sigma Aldrich. Commercial sunflower oil was used. Food grade ingredients were always used.

2.2 Interfacial Properties

Surface dilatational measurements were carried out at 20 °C using an automatic pendant drop tensiometer (TRACKER, IT Concept, France). The surface dilatational modulus (E) was obtained through a periodic sinusoidal interfacial compression and expansion performed by decreasing and increasing the drop volume at 10% of deformation amplitude ($\Delta A/A$) and 100 MHz of angular frequency (ω). The sinusoidal oscillation for surface dilatational measurement was made with 5 oscillation cycles followed by a time of 50 cycles without any oscillation up to the time required to complete adsorption.

2.3 Rheological Properties

An AR 2000 rheometer (TA Instruments, USA) was used, with a 60 mm (in diameter) parallel plate low inertia geometry and a gap size of 0.5 mm, being shear rate increased step-by-step from 1 to 100 s^{-1}, obtaining a steady state for each value.

3 Results and Discussion

3.1 Dilatational Tests

Figure 1 shows the values for the interfacial dilatational modulus, E, and loss tangent, tan δ, along time for systems containing β-lactoglobulin in the absence and in the presence of 10% wt. inulin. The presence of inulin leads to a reduction in both parameters, which may be related to the increased adsorption rate onto the O/W interface when inulin is present. This would lead to an accumulation of protein in the interface in a relatively short period of time, preventing the formation of a coherent film, and resulting in a reduction of interactions between protein-protein molecules in the interface.

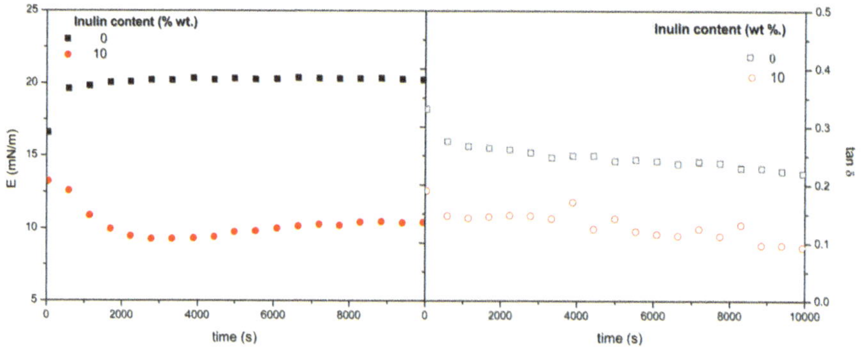

Fig. 1. Dilatational elastic modulus (left) and loss tangent (right) evolution for β-lactoglobulin, 0.5% wt., in the absence and in the presence of inulin (10% wt.)

3.2 Flow Curve Tests

Figure 2 shows the flow curves in terms of viscosity vs shear rate for the same emulsion systems, 1 and 15 days after their preparation. It may be observed how inulin results in a higher viscosity, also acquiring a shear-thinning behaviour, which is related to the thickening effect of the polysaccharide. On the other hand, the stability of these emulsions is verified through the lack of evolution of the viscosity along time, as all flow curves remain virtually the same after 15 days since the preparation of the emulsions.

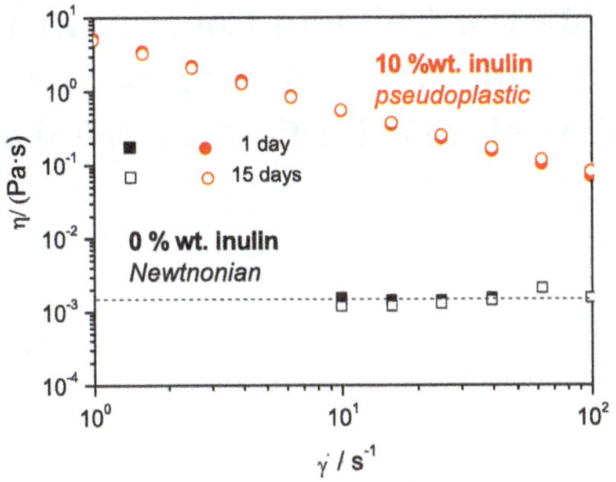

Fig. 2. Viscosity as a function of shear rate for emulsions stabilized by β-lactoglobulin, 0.5% wt., in the absence and in the presence of inulin (10% wt.)

4 Conclusions

These results support the use of a prebiotic ingredient like inulin in commercial food emulsions, as stable emulsions are formed. Moreover, its role as thickener and the importance of the interactions with the emulsifier used has been highlighted.

References

1. Dickinson, E.: Hydrocolloids at interfaces and the influence on the properties of dispersed systems. Food Hydrocoll. **17**, 25–39 (2003)
2. Korbelik, M., Cooper, P.D.: Potentiation of photodynamic therapy of cancer by complement: the effect of γ-inulin. Br. J. Cancer **96**, 67–72 (2007)
3. Adebola, O.O., Corcoran, O., Morgan, W.A.: Synbiotics: the impact of potential prebiotics inulin, lactulose and lactobionic acid on the survival and growth of lactobacilli probiotics. J. Funct. Foods **10**, 75–84 (2014)

Effect of Chitosan Concentration on the Rheological Properties of Acetic and Lactic Acid Solutions

Clara A. Tovar[1(✉)], M. Carmen Gómez-Guillén[2], and M. Pilar Montero[2]

[1] Department of Applied Physics, Faculty of Sciences, University of Vigo, As Lagoas s/n, 32004 Ourense, Spain
tovar@uvigo.es

[2] Department of Products, Institute of Food Science, Technology and Nutrition (ICTAN-CSIC), C/José Antonio Novais, 10, 28040 Madrid, Spain
{cgomez,mpmontero}@ictan.csic.es

Abstract. Chitosan (Ch) is a biodegradable cationic polymer, with industrial interest. The purpose of this study was to analyse the effect of Ch concentration (1–3%), on the rheological properties of acetic (A1–A3) and lactic (L1–L3) acid solutions. Flow curves (up-down) at 20 °C, provided the consistency (κ) and the flow index (n). In all solutions, κ increased and n decreased with Ch concentration. For L3, greater κ value and lower n exponent indicate stronger shear thinning response. The structural-breaking rate (v_{up}), and the regeneration rate (v_{down}) were proposed as a measurement of the time dimension in the shear thinning response.

1 Introduction

Chitosan (Ch) is a non-toxic cationic polymer, which has been received interest in the pharmaceutical and food industry because of its functional properties [1] which are dependent on the pH and type of acid in which it is dissolved. The replacement of acetic by lactic acid for Ch solutions may be advantageous to avoid the odour associated to the former, however, it might modify the rheological behaviour. It has been observed that in the preparation of Ch films, the addition of plasticizing agents is not necessary when the Ch is solubilized in lactic acid, which can constitute an extra benefit [2]. The objective of this work was to evaluate: (1) the effect of the (lactic and acetic) acid solutions on the flow response at several Ch concentrations. (2) The potential applications of two acids as a function of Ch concentration.

2 Materials and Methods

Chitosan (deacetylation degree 90.7%; molecular weight 140,000 Da) from Guinama (Valencia, Spain), was dissolved in acetic acid solutions at 1% (w/v) (A1, pH = 3.50), 2% (w/v) (A2, pH = 3.79) and 3% (w/v) (A3, pH = 3.98) Ch concentration. Similarly

F. J. Galindo-Rosales et al. (Eds.): IBEREO 2019, SPM, pp. 20–24, 2020.
https://doi.org/10.1007/978-3-030-27701-7_5

with 1% (w/v) (L1, pH = 3.20), 2% (w/v) (L2, pH = 4.06) and 3% (w/v) (L3, pH = 5.52) Ch in lactic acid solutions under stirring overnight at room temperature (20 °C).

Rheological tests were performed with RS600 Haake rheometer (Thermo Electron Corp. Karlsruhe, Germany) and the cone-plate (CP35/2 with 0.105 mm gap). All measurements were made at 20.0 ± 0.1 °C. The viscoelastic properties were analysed by time sweeps (5400 s, 0.1 Hz and $\gamma = 0.4\%$). Flow curves were obtained using four intervals: (1) pre-shear phase (30 s^{-1}, 10 min); (2) up ramp (300→900 s^{-1}; 10 min.); (3) high rate interval (900 s^{-1}, 5 min.); (4) down ramp (900→300 s^{-1}; 10 min). The thixotropy was examined by the step test with three steps: (1) reference (100 s^{-1}, 90 s); (2) high shear-rate (1000 s^{-1}, 45 s); (3) regeneration (100 s^{-1}, 600 s). Statistical analysis: data are presented as mean values of five replicates and were tested with expanded uncertainty limits (EUL). Trends were considered significant when means of compared sets differed at $p < 0.05$ (Student's t-test).

3 Results and Discussion

3.1 Time Dependence of the Oscillatory Measurements

G' and G'' were practically time independent and data were means of the instantaneous values considering the testing times and the number of test repetitions. All solutions exhibited fluid-like behaviour with phase angle ($\delta > 50°$). At fixed Ch concentration, lactic acid solutions exhibited significantly higher G' and G'' and lower δ values ($p < 0.05$) than in acetic solution (Table 1), showing a more "fluid strength". This result shows the greater stabilising role of lactic acid with Ch chains, principally by hydrogen bonding and polar interactions, in line with the plasticizer effect shown of Ch films [2].

Table 1. Viscoelastic parameters (Pa) of different acid solutions at 0.1 Hz and 20 °C

	A1	L1	A2	L2	A3	L3
G'	0.10 ± 0.01^{b}	0.17 ± 0.03^{c}	0.0173 ± 0.003^{a}	0.53 ± 0.09^{e}	0.38 ± 0.05^{d}	0.93 ± 0.14^{f}
G''	0.15 ± 0.01^{a}	0.25 ± 0.04^{b}	0.164 ± 0.004^{a}	0.97 ± 0.12^{c}	1.11 ± 0.08^{d}	2.11 ± 0.09^{e}
δ (°)	59.0 ± 2.7^{b}	53.0 ± 3.2^{a}	84.25 ± 0.79^{e}	65.5 ± 2.8^{c}	73.5 ± 1.4^{d}	68.9 ± 2.4^{c}

$^{a-f}$ different letters in the same row indicate significant differences ($p < 0.05$) for each parameter

The greater differences were found at 2% Ch (Table 1), suggesting that L2 solution is a more complex and structured fluid comparing with A2. For L3, a similar result was found, but naturally with higher G' and G'' moduli (Table 1). At higher Ch concentration, there are more entanglements producing a denser fluid.

3.2 Flow Curves

Flow curves were shown in terms of the apparent viscosity (η) *vs* shear rate ($\dot{\gamma}$) (Eq. 1).

$$\eta = \kappa \cdot \dot{\gamma}^{n-1} \tag{1}$$

Being κ the consistency, and n the flow index [3]. κ values increased with increasing Ch concentration, being similar in up and down curves, except for L3, since down-κ > up-κ (Fig. 1a). This result indicates greater structural regeneration ability with decreasing shear rate (L3).

At fixed Ch concentration (2% and 3%), the κ coefficient was higher in L2 *vs* A2 and in L3 *vs* A3, showing a greater packing level in lactic *vs* acetic acid solutions. In A1 and L1 n parameter was around 0.93 (Newtonian response), since η was practically shear independent. At higher Ch concentrations, n values decreased being the lowest in L3. In addition, down-n < up-n (Fig. 1b), low n values indicate a shear thinning behaviour. So, when the shear rate increases, the viscosity decreases, since the molecular friction between fluid layers decreases, thus, the Ch chains might be aligned in the shear direction, reducing the flow resistance [3].

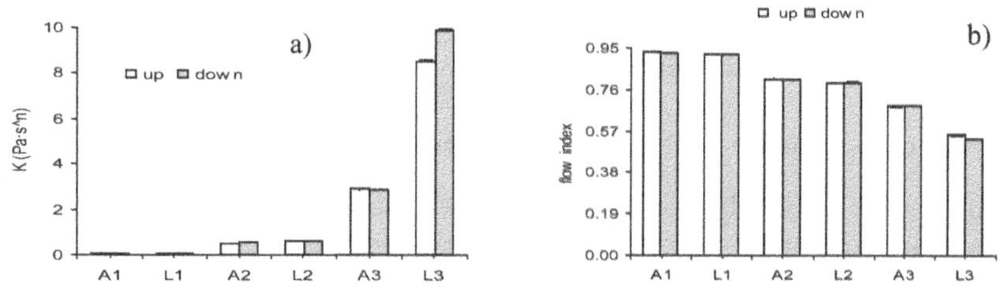

Fig. 1. Influence of the chitosan concentration in acetic and lactic acid solutions on the consistency (κ) (a) and flow index (n) (b). T = 20 °C

In general η values can be linearly fitted with time (Eq. 2). From the absolute value of the slope (b) (up and down curves), can be obtained the structural disruption and regeneration rate respectively (Table 2). Thus, b slope quantifies the temporal dimension of the "shear-thinning", phenomenon dependent on the colloidal structures [3]. So, a complex fluid is a more shear sensible and η is faster reduced (up-curve) being a shear-reversible effect due to the physical nature of the intermolecular interactions. Thus, in the down-curve new structural rearrangements in "*chitosan- acids-water*" complex produced a specific re-construction of the inner structure [4] based on the type of acid and Ch concentration.

$$\eta = \eta_0 + b \cdot t \qquad (2)$$

Table 2. Rate of the structural disruption (up) and regeneration (down) for A1–A3 and L1–L3 solutions. T = 20 °C

Sample	A1	L1	A2	L2	A3	L3
Up-b (μPa)	-4.415 ± 0.004^{aA}	-4.81 ± 0.01^{aB}	-49.6 ± 0.4^{aC}	-57.4 ± 0.5^{aD}	-212.87 ± 3.6^{aE}	b + 2ct (μPa) up −663 + 0.93t
Down-b (μPa)	4.666 ± 0.008^{bA}	5.01 ± 0.02^{bB}	50.6 ± 0.4^{bC}	57.3 ± 0.6^{aD}	210.1 ± 3.0^{aE}	b + 2ct (μPa) down 125 + 0.95t

[a–b] different small letters indicate significant differences ($p < 0.05$) between up and down curves; [A–E] different capital letters indicate significant differences ($p < 0.05$) between A1–A3 and L1–L2 samples.

In acetic acid solutions the *b* slopes were similar in up *vs* down ramps. The *b*-slope was significantly ($p < 0.05$) higher in A3→A2→A1. Similarly occurs in L1 and L2 (Table 2). At fixed Ch concentration (1% and 2%), *b* values were lower in acetic than in lactic acid solutions (Table 2). This fact could be related to the greater steric impediments in the acetic solutions due to the methyl groups in shorter molecular chains (A1 and A2). The repulsive forces could reduce the "*fluid strength*" in A1 *vs* L1 and in A2 *vs* L2 (Table 1). L3 exhibited a quadratic dependence of η *vs* t (Eq. 3). Hence, the first derivative is a time-linear function whose down-slope (rate of structural regeneration) was higher than the up-slope (structural disruption) (Table 2).

$$\eta = \eta_0 + b \cdot t + c \cdot t^2 \tag{3}$$

This trend suggests that L3 exhibited higher number of the associating sites in the recovery process, comparing with those at the initial state. This fact could be related to the bioadhesive role in the pharmaceutical industry.

3.3 Step Test with Three Intervals

The possible thixotropy was examined using the step test with three intervals. For all solutions viscosities at constant shear rate for each step were time independent, irrespective of the shear rate value. Hence, there was not thixotropy in any case (Fig. 2). It might be calculated the percentages of the structural regeneration (SR), considering the relative increase of the viscosity, from η(t = 135 s) to η(t = 750 s) (Eq. 4). For 1% Ch, %SR were 12% (A1) and 14% (L1). This percentage was increased with increasing the Ch concentration, being similar in A2 (40%) *vs* L2 (44%). However at 3% Ch, there were differences between %SR in A3 (88%) and L3 (280%), in line with the greater b-slopes (Table 2). So, lactic acid promotes the formation of greater amount of intermolecular hydrogen bonds without steric effects enhancing the molecular reconstruction ability in L3 solution.

$$\%SR = \frac{(\eta_{750} - \eta_{135}) \cdot 100}{\eta_{135}} \tag{4}$$

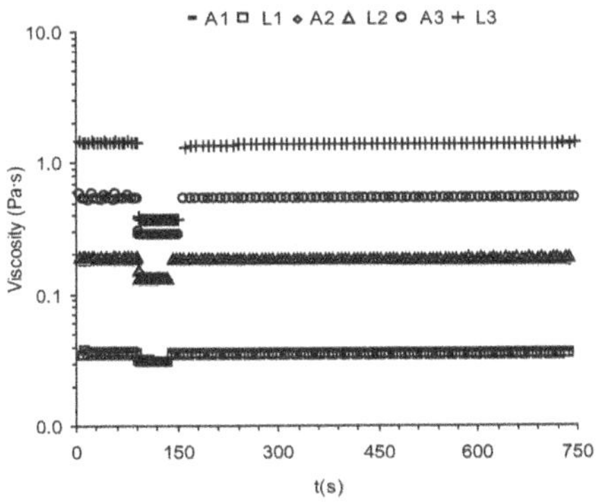

Fig. 2. Influence of the chitosan concentration and type of acid on the viscosity (100 s^{-1}, 1000 s^{-1} and 100 s^{-1}). T = 20 °C

Acknowledgments. This work has been financed by the Spanish MINECO under Project AGL2017-84161 co-funded with EU-Regional Development Fund.

References

1. Cheung, R.C.F., Ng, T.B., Wong, J.H., Chan, W.Y.: Mar. Drugs **8**, 5156–5186 (2015)
2. Arancibia, M., Alemán, A., Calvo, M.M., López-Caballero, M.E., Montero, P., Gómez-Guillén, M.C.: Food Hydrocoll. **23**, 710–717 (2014)
3. Figura, L.O., Teixeira, A.A.: Food Physics, pp. 149–152. Springer, Heidelberg (2007)
4. Larson, R.G.: The Structure and Rheology of Complex Fluids, pp. 12–15. Oxford University Press, New York (1999)

Rheological Methods to Evaluate the Efficacy of a New Compounding to Diagnose the Olfactory Dysfunction

Carolina Chaves[1], Joana Marto[1], Filipa Duarte-Ramos[1], Armando Alcobia[2], Maria Rosário Bronze[1,3], and Helena Margarida Ribeiro[1(✉)]

[1] Research Institute for Medicines (iMed.ULisboa), Faculty of Pharmacy, Universidade de Lisboa, Lisbon, Portugal
carolina.tovar.chaves@gmail.com, {jmmarto, f.duarteramos,mrbronze}@ff.ulisboa.pt, hribeiro@campus.ul.pt

[2] Hospital Garcia de Orta, EPE, Almada, Portugal
aalcobia@hgo.min-saude.pt

[3] iBET, Apartado 12, 2781-901 Oeiras, Portugal

Abstract. The assessment of olfactory dysfunction (partial or total anosmia) is important in the early diagnosis of neurodegenerative diseases and may be a good indicator in clinical diagnosis. Currently, there is no diagnostic test for partial or total anosmia validated for the Portuguese population. The aim of this study was the development and rheological and sensory characterization of hospital compounded formulations containing 23 different fragrances adapted to Portuguese population. During the studies, several PEG-based formulations were developed. Viscoelasticity assays were performed in the Rheometer Kinexus using the oscillatory method. Differential scanning calorimeter and microscopy were assessed as well as the volatile components analysis by GC-MS. Thus, a new compounding for the diagnosis of olfactive dysfunction with a structured and rheological characteristic adapted to the Portuguese population were developed.

1 Introduction

The assessment of olfactory dysfunction (partial or total anosmia) is important in early diagnosis of neurodegenerative diseases and may be a good indicator in clinical diagnosis. The olfactory threshold, defined as the minimum measure detectable for a given odor, increases with age (1% per year), being this effect less significant in women than in men. In the elderly, the decline of this sense occurs more marked and with age there is also a decrease in the ability to discriminate the sensory characteristics that relate to the sense of smell. This olfactory decrease (diosmia) occurs due to the physiological process of aging, occurring in the sixth-seventh decade, but also occurs in neurodegenerative diseases such as Alzheimer and Parkinson [1]. Olfactory function might be evaluated by different methodologies such as sensory methods, electro-physiological or neuro imaging techniques. Currently, there is no diagnostic test for

F. J. Galindo-Rosales et al. (Eds.): IBEREO 2019, SPM, pp. 25–29, 2020.
https://doi.org/10.1007/978-3-030-27701-7_6

partial or total anosmia validated for the Portuguese population, nor any adapted test, as is already the case in other European populations. The most frequently used device in Portugal, the Barcelona Smell test-24, has some smells that our population does not recognize, as they are not part of our cultural habits, what may introduce difficulties in test application and lead to misdiagnosis.

The aim of this study was the development and rheological and sensory characterization of hospital compounded formulations containing 23 different fragrances adapted to Portuguese population.

2 Materials and Methods

2.1 Materials

Polyethylene glycol 1500 was purchased from José M. Vaz Pereira, S.A., Portugal and Polyethylene glycol 400 was obtained from Sigma, Germany. Fragrances were obtained from Aromol, Portugal.

2.2 Methods

2.2.1 Preparation of the Formulation

During the studies, several PEG-based formulations were developed. The effect of parameters such as the odor and percentage in PEGs (PEG-4000, PEG-1500, PEG-600, PEG-400) on formulation development were assessed.

2.2.2 Macroscopic and Microscopic Observation

The structure of the five final formulations were analyzed by Optical Microscopy. A computerized image analysis device was used for the microscopic observation connected to an Olympus BX40 microscope in bright field (Olympus, Japan).

2.2.3 Rheological Measurements

Viscoelasticity assays were performed in the Rheometer Kinexus (Malvern) using cone and plate geometry (truncated cone angle 4° and radius 40 mm). All oscillatory tests were performed at 0.001% oscillation strain, which is well within the linear viscoelastic range of the formulations. The frequency sweep test was performed over a frequency range from 1 to 40 Hz, at a shear stress of 0.001 Pa. All tests were performed on samples of about 1 g, at 25 ± 1 °C.

2.2.4 Differential Scanning Calorimetry

The thermos analytical measurements were performed with a calorimeter DSC Q-200 model (TA Instruments®, USA). Samples were weighed on aluminium pans which were then crimped with a lid on. The pan was equilibrated at 40 °C followed by heating at 10 °C/min up to 120 °C.

2.2.5 Gas Chromatography–Mass Spectrometry (GC-MS)

Analysis of volatile compounds was carried out in gas chromatography (GC) associated with mass spectrometry (MS) with autosampler. The GC–MS equipment consisted on a

QP 2010 Plus, Shimadzu equipped with ZB-5MSi column (5%-Phenyl-95% Dimethyl polysiloxane column 30 m x 0.25 mm i.d., film thickness 0.25 µm; Zebron GC columns, Phenomenex, and an AOC-5000 Shimadzu autosampler.

3 Results and Discussion

The macroscopic observation results showed that all the formulations were uniform, had light yellow color, and did not present any incompatibility in terms of PEGs (Fig. 1).

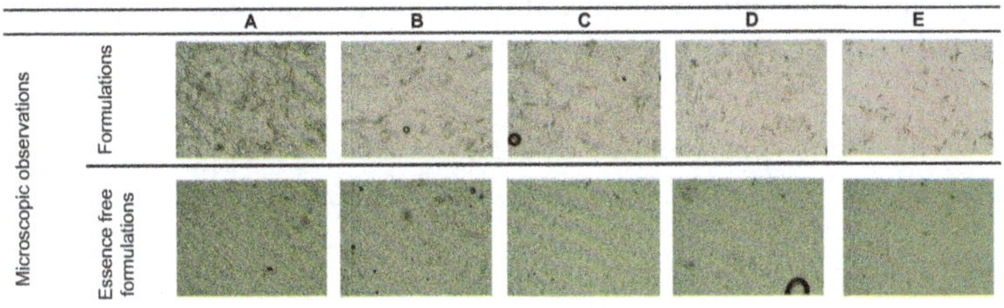

	A	B	C	D	E
PEG 1500 (%)	60	50	40	00	20
PEG 400 (%)	20	30	40	50	60
Essence (%)	20	20	20	20	20

Fig. 1. Composition and microscopic images of different formulations (Magnification: 100x).

There is no difference between A and E in the number of crystals because the percentage of essence is the same. As the amount of PEG-1500 increases the structure becomes more complex, which eludes an image apparently with more crystals, which is not true. PEG-1500 has a network structure that is more evident in A.

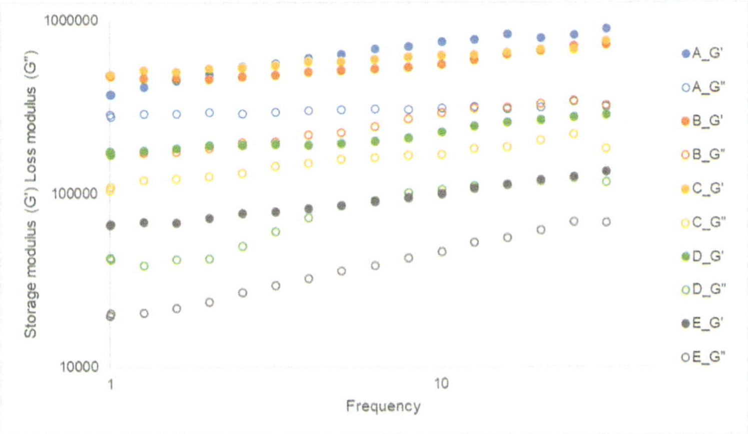

Fig. 2. Frequency sweep results of formulations A–E.

In all formulations, the storage modulus > loss modulus meaning these formultions have the elastic module superior to the viscous module, therefore presenting a strong network that difficult the flow and consequently the filling process. Hardness and cohesion are related to elasticity (Fig. 2). Rheological studies on the flow properties of PEG indicated that both modules, the elastic and the viscous, increased as the percentage of PEG-1500 increased as indicated in multiple studies. Formulations containing PEG-400 and PEG-1500 obtained G′ < G″, when compared to formulations containing mixtures of PEG-400, PEG-1500 and PEG-6000. Elastic and viscous modules values also depend on the fragrances used in the formulations [2].

DSC thermograms showed that formulations containing PEG-400 and PEG-1500 presented a shift of the PEG-1500 endothermic peak (from 46.5 °C to 37 °C) which was expected.

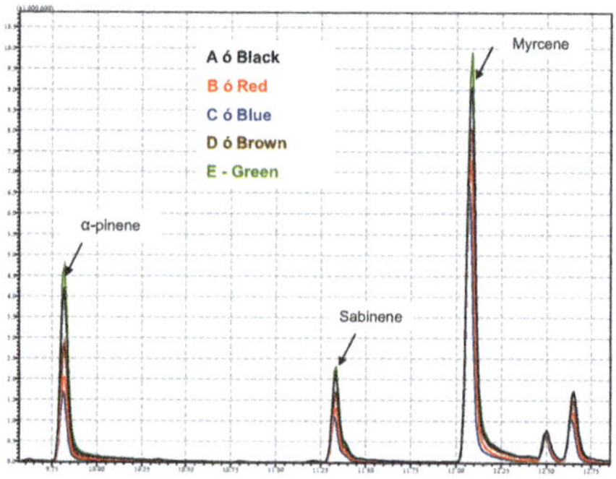

Fig. 3. Part of chromatograms obtained by SPME-GC-MS corresponding to A–E formulations.

The chromatograms (Fig. 3) of the C and D formulations are those presenting the smaller peak areas for α-pinene and myrcene compounds. This reinforces the idea that having a larger peak area does not always mean better sensory acceptance. The most adequate formulation was 50:50 of PEG-1500:PEG-400.

4 Conclusions

Our findings uphold the formulations development importance regarding its rheological and sensorial characteristics adapted to the Portuguese population to improve the diagnosis of olfactive dysfunction. A formulation for the diagnosis of olfactive dysfunction adapted to the Portuguese population was developed as reliable diagnostic method for total or partial anosmia.

References

1. Lawton, M., Hu, M.T., Baig, F., et al.: Equating scores of the University of Pennsylvania Smell Identification Test and Sniffin' Sticks test in patients with Parkinson's disease. Parkinsonism Relat. Disord. **33**, 96–101 (2016)
2. Ribeiro, H.M., Morais, J.A., Eccleston, G.M.: Structure and rheology of semisolid O/W creams containing cetyl alcohol/non-ionic surfactant mixed emulsifier and different polymers. Int. J. Cosmeti. Sci. **26**, 47–59 (2004)

Ocular Lubricants Efficacy: Mucoadhesive Evaluation Using Rheological Methods

Angélica Graça[1], Lídia Maria Gonçalves[2], Sara Raposo[1], Helena Margarida Ribeiro[2], and Joana Marto[2(✉)]

[1] Laboratório Edol–Produtos Farmacêuticos, S.A., Linda-a-Velha, Portugal
{angelicagraca, sraposo}@edol.pt
[2] Faculty of Pharmacy, Research Institute for Medicine (iMed.ULisboa), Universidade de Lisboa, Lisbon, Portugal
{lgoncalves, jmmarto}@ff.ulisboa.pt, hribeiro@campus.ul.pt

Abstract. Polymer-based eye drops are the most drug delivery system used to treat the dry eye disease (DED). Therefore, the mucoadhesion between the polymer and the ocular mucin is crucial to ensure the efficacy of the treatment. In this context, the present study aimed to evaluate the potential use of *in vitro* methods to study the mucoadhesion of two Hyaluronic Acid (HA)-based eye drops solutions, HA 0.15% and 0.3% (w/v) to treat DED. Two rheological methods were performed to study the mucoadhesive properties of both formulations: tackiness testing and oscillation frequency sweep. *In vitro* tests on ARPE-19 cell line were performed using a 2D and a 3D dry eye model. The results showed HA 0.3% appeared to be tackier than HA 0.15% in both cases. *In vitro* tests on ARPE-19 have shown that pre-treated cells with HA showed a morphology more similar to the hydrated cells in both products, with a high survival rate. The *in vitro* techniques used in this study have shown to be suitable to evaluate and predict mucoadhesive properties and the efficacy of the eye drops on relief or treat the DED. The results obtained from these methods may help in inferring possible *in vivo* effects.

1 Introduction

Dry eye disease (DED), also known as keratoconjunctivitis sicca, is a pathology whose origin comes from several factors, resulting in symptoms of discomfort, visual disturbance, tear film instability with potential damage to the ocular surface [1]. Lubricants eye drops are the most used form of treatment due to its simple administration and immediate relief. The presence of polymers is crucial to improve mucoadhesive properties of artificial tears [2]. Hyaluronic Acid (HA) has been used with success in treating patients with severe DED. To understand the eye drop efficacy the study of the mucoadhesion is of great importance. The most common and suitable methods to assess the mucoadhesive properties of a potential formulation candidate for ocular delivery is through *in vitro* techniques. *In vitro* cell viability and morphology tests or ARPE-19 cell lines are also useful techniques to evaluate the efficacy of the product [3]. Thus, the aim of this research work was to study the mucoadhesivity of two eye

F. J. Galindo-Rosales et al. (Eds.): IBEREO 2019, SPM, pp. 30–34, 2020.
https://doi.org/10.1007/978-3-030-27701-7_7

drops formulations containing HA, HA 0.15% and HA 0.30%, and to validate *in vitro* methods to be used in the future to study mucoadhesion properties in detail. A compilation of several rheological methods was performed, namely tackiness testing and oscillation frequency sweep. The zeta potential (ZP) was also studied. As a complementary study of the product efficacy, a cell viability assay was performed using 2D and 3D culture cells models that mimic the conditions given by DED (Fig. 1).

2 Materials and Methods

2.1 Materials

High molecular weight sodium hyaluronate was a kind gift from Inquiaroma, Spain. It was use dried mucin from porcine stomach type II (Sigma-Aldrich, USA). Pig eyes were obtained from a local slaughterhouse. Human retinal pigment epithelial cell lines ARPE-19 (ATCC® CRL-2302™) were obtained from American Type Cell Culture collection (USA), and they were used for cell viability and dry eye assays. Cell culture medium and supplements were from Gibco (Thermo Fisher Scientific, UK).

2.2 Mucoadhesive Studies

The oscillatory tests were performed using a Malvern Kinexus Rheometer (Malvern Instruments, Malvern, UK) with cone and plate geometry (truncated cone angle 4° and radius 40 mm). In order to determine the linear viscoelastic range (LVER), the oscillation amplitude test was performed at a frequency of 1 Hz, shear strain was varied by increasing the oscillation amplitude from 0.01% to 100%. In the present study, we conducted all oscillatory tests only at 0.25% oscillation strain, which corresponds to the LVER of the formulations. In the frequency testing the frequency range used was between 10–0.1 Hz. The adhesive strength was also measured using the same equipment and a plate and plate geometry. It was used a toolkit with the conditions of 0.1 mm/s, 5 mm and 0.15 GAP. The same protocol was performed using pig eyes obtained from a local slaughterhouse, instead of mucin. The mucoadhesion interaction was also determined by measuring the zeta potential (ZP) of the mixtures of mucin and each solution using a Zetasizer Nanoseries Nano Z (Malvern Instruments, Malvern, UK).

2.3 2D and 3D Culture Cells Models

The protective effect of the selected formulas against dehydration was evaluated using previously reported protocols [3]. A 2D and 3D *in vitro* assay was performed to study the differences in the morphology and cell viability of dehydrated cells who received a pre-treatment with HA 0.15% and HA 0.30% formulations with cells who were not submitted to the treatment. It was also tested the commercial formulation (CR) with 0.30% of HA.

3 Results and Discussion

Two eye drop solutions with different HA concentrations were developed, HA 0.15% and HA 0.30% (w/v). Both products are isotonic, with a limpid and clear aspect and a pH value between 7.0–7.6, which are similar to the lacrimal fluid to avoid eye irritation and provide ocular lubrication and comfort.

3.1 Oscillatory Measurements

In both products the G" is greater than G' at low frequencies, which indicates a fluid-like system. With the addition of mucin this profile still remains but the values of both shear moduli increase. This means that it is necessary a greater amount of force or stress to deform the sample along the plane of the direction of the force, which indicates that some type of interaction has been established. The interaction between a mucoadhesive polymer with mucin may occur by the following mechanisms: physical entanglements, Van der Walls bonds, electrostatic forces and hydrogen bonds. The shear moduli present greater values in HA 0.30% in comparison with HA 0.15% which indicates that the strength of the formulation/mucin interaction increases with HA concentration.

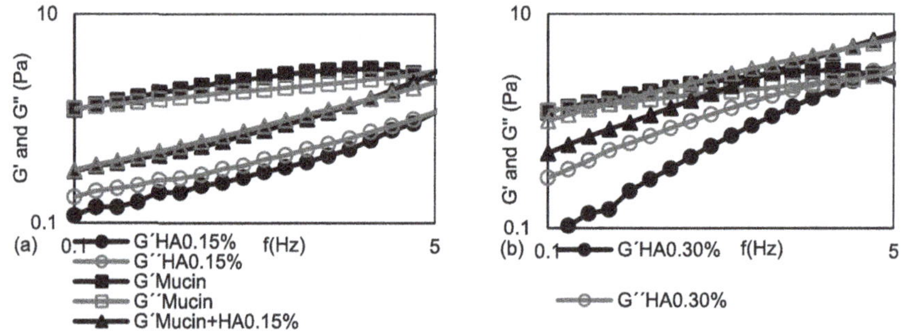

Fig. 1. Frequency sweep with shear moduli as function of frequency of HA 0.15%, Mucin 5% and Mucin + HA 0.15% (a) and of HA 0.30%, Mucin 5% and Mucin + HA 0.30% (b) at room temperature.

3.2 Tackiness Testing

The significant difference between HA 0.15% vs Mucin + HA 0.15% and HA 0.30% vs Mucin + HA 0.30% (Table 1) is an indication that the addition of mucin created an interference with the polymer which formed a more viscous system. A similar study was performed but instead of mucin was used pig eye. Three samples were used and attached to the probe and the adhesive force between the eye and samples was measured. The results showed there are significant differences between HA 0.15% and HA 0.30%, were the HA 0.30% appears to be the tackiest and the strongest. The strengthening might arise from the entanglement of the polymer chains and the mucus glycoproteins, the formation of chemical bonds and/or from dehydration of the mucus layer.

Table 1. Normal force and area under force time curve results for HA 0.15%, HA 0.30%, CR 0.15%, CR 0.30%, Mucin, Mucin + HA 0.15% and Mucin + HA 0.30% (Mean ± SD, n = 6).

	Peak normal force (N)	Area under force time curve (N.s)
HA 0.15%*1	−0.18 ± 0.00	0.44 ± 0.06
HA 0.30%*1	−0.23 ± 0.01	0.77 ± 0.09
CR 0.15%*1	−0.17 ± 0.02	0.90 ± 0.07
CR 0.30%*1	−0.22 ± 0.01	1.05 ± 0.04
Mucin*1	−0.23 ± 0.00	1.01 ± 0.06
Mucin + HA 0.15%*1	−0.22 ± 0.02	0.57 ± 0.15
Mucin + HA 0.30%*1	−0.29 ± 0.03	0.75 ± 0.07
Pig Eye + HA 0.15%*2	−0.08 ± 0.03	0.89 ± 0.06
Pig Eye + HA 0.30%*2	−0.13 ± 0.03	1.01 ± 0.06

*1Mean ± SD, n = 6; *2Mean ± SD, three different eyes, n = 3; CR- Commercial Reference.

3.3 Zeta Potential

The ZP study demonstrated the values of HA 0.15% and HA 0.30% are similar to their market equivalent formulation, CR 0.15% and CR 0.30%, respectively. However, comparing the values of the two products HA 0.15% ZP values are much more negative that HA 0.30% values. These negative values are in accordance to the fact the HA presents an anionic nature due to the presence of carboxylic groups. Regarding the viscosity, in negatively charged systems repulsive forces are greater, which is also an indicator of fluid-like properties [3]. The mucin also presents negative charge due to the oligosaccharide chains which confer negative charge to the mucins through carboxyl and sulphate groups. When the mucin is added to both products an increase of the negative charge is observed, being the ZP value more negative in Mucin 5% + HA 0.30% than with Mucin 5% + HA 0.15%.

3.4 Cell Morphology and Cell Viability – 2D and 3D Model

The results have shown that the cells that were not treated with HA presented a disintegrated and dry membrane with a cell mortality rate close do 50% when compared to the non-dehydrated cells. The pre-treated cells showed a morphology more similar to the hydrated cells in both products with a high survival rate (62.7% ± 9.5% with HA 0.15% and 72.5% ± 6.2% with HA 0.30%). In the 3D model the results show that the application of all three formulations obtained over 100% cell viability meaning that the application of these MD provided a more suitable environment for cell proliferation. The reason may be due to the protective effect displayed by the HA formulation on the cells is related to the polymer water retaining capacity. HA 0.30% presented higher cell viability values since more concentrated formulations retain more water promoting higher hydration.

4 Conclusions

In conclusion, rheological methods have proved that the viscosity increased with the increase of HA concentration and the interactions occurred between mucin and HA are due through possibly entanglements and hydrogen bonding. In the 2D and 3D models it was concluded that the cells pre-treated with HA preserved the cell's morphology after the dehydration process and maintained a high survival rate. From these results obtained the chosen *in vitro* methodology as well as, cell viability assay, demonstrated to be suitable and useful to study mucoadhesivity.

References

1. Cwiklik, L.: Tear film lipid layer: a molecular level view. BBA Biomembranes **1858**(10), 2421–2430 (2016). https://doi.org/10.1016/j.bbamem.2016.02.020
2. Ludwig, A.: The use of mucoadhesive polymers in ocular drug delivery. Adv. Drug Deliv. Rev. **57**(11), 1595–1639 (2005). https://doi.org/10.1016/j.addr.2005.07.005
3. Graça, A., Gonçalves, L., Raposo, S., Ribeiro, H.M., Marto, J.: Useful *in vitro* techniques to evaluate the mucoadhesive properties of hyaluronic acid-based ocular delivery systems. Pharmaceutics **10**, 110 (2018). https://doi.org/10.3390/pharmaceutics10030110

Rheological Tools Used in the Development of an Oral Vehicle for Paediatric Patients

Margarida Pereira[1], Filipa Cosme Silva[1,2], Helena Margarida Ribeiro[1], António José Almeida[1], and Joana Marto[1(✉)]

[1] iMed.ULisboa, Faculty of Pharmacy, Universidade de Lisboa, Lisbon, Portugal
margaridapereira95@hotmail.com, cosme.filipa@gmail.com, hribeiro@camous.ul.pt, {aalmeida, jmmarto}@ff.ulisboa.pt
[2] Hospital de Santa Maria, Centro Hospitalar Lisboa Norte, EPE, Lisbon, Portugal

Abstract. Palatability in the paediatric population is perceived as a true challenge once their ability to take solid oral dosage forms is far more reduced than adults, which often can lead to a poor therapeutic compliance. In Portugal, there is not any vehicle to aid the administration of drugs to paediatric patients so, the aim of this project was to develop a hydrogel vehicle to facilitate the intake of tablets, capsules and other types of oral medicines in children. This research work allowed us to demonstrate how rheological methodologies were crucial to optimize some physical properties of the vehicle, thought to be critical to improve the patient's experience during the oral intake of a drug.

1 Introduction

Formulation acceptability and preferences facilitate medication paediatric compliance, and they constitute important factors in achieving the intended treatment outcomes. Paediatric drug administration can be very challenging as they are usually more sensible to flavours, can have swallowing difficulties and may have a poor pharmaceutical adherence. Many existing formulations are not suitable for children, which often leads to off-label and unlicensed use of adult medicines. Most of times, physicians, nurses and caregivers have to find new strategies to overcome difficulties in the administration of the medicine, including crushing tablets and opening of capsules [1]. To minimise unnecessary manipulation of drugs, further work shall be done to identify globally acceptable and available vehicles that should be the first option for co-administration with oral dosage forms. Taking this into account, the main goal of this project was the development of a hydrogel vehicle to facilitate the intake of tablets, capsules and other types of oral medicines in children. On Portugal there is not any product with these characteristics on the market. On Europe, there is a product named Gloup but it contains sugar which can be a concern for a chronic use. That is the main reason why the authors of this work believe that it was important to develop a new vehicle, without

F. J. Galindo-Rosales et al. (Eds.): IBEREO 2019, SPM, pp. 35–39, 2020.
https://doi.org/10.1007/978-3-030-27701-7_8

sugar, with adequate excipients for paediatric use, with a color that can mask the powder or the tablet to be administrated and with appropriate reology, texture and taste.

Thus, the already existing product Gloup® was used as a reference and a series of polymers thought to have similar proprieties were chosen. In order to evaluate which polymer came the closest to our reference, three rheological methods were used – viscosity, oscillation frequency and mucoadhesion tests.

2 Materials and Methods

2.1 Raw Materials

The polymers used in this study were: sodium carboxymethylcellulose (CMC) and methylcellulose (MC) from Fagron (Spain), Natrosol® - hydroxyethylcellulose (HEC) from Aqualon (Spain), gelatine and xanthan gum from Disproquima (Portugal), and hyaluronic acid 1.8-2.2Eyed™ (HA) from Inquiaroma (Spain). Other excipients were used, like sodium methylparaben, sodium propylparaben and sodium citrate obtained from Fagron (Spain), and sucralose and citric acid monohydrate from António M. S. Cruz (Portugal).

2.2 Procedures

The rheological characterization of the hydrogels was achieved by using continuous shear and oscillation techniques. The experiments were performed with a controlled stress Kinexus Rheometer (Malvern Instruments, Worcestershire, UK) using cone and plate geometry (truncate angle 4° and radius of 40 mm). The viscosity test used a range of frequencies from 0.1 to 100 Hz, at 25 °C. The frequency sweep method was performed between 0.1 and 10 Hz, with a shear strain of 3.5%, at 25 °C. The mucoadhesion was also measured using the same equipment and a plate-plate geometry. The parameters used were 0.1 mm/s for the gapping speed, 5 mm for the final gap and 0.2 mm for the working gap.

3 Results and Discussion

3.1 Formulation Development

A total of 15 formulations were developed and the concentration of each polymer was adjusted in order to obtain the desired characteristics on the final vehicle. The concentration of the various polymers ranged from 1.5% to 10%. The composition of the hydrogels is described in Table 1.

Table 1. Quantitative and qualitative composition of the formulations.

Excipients (chemical name)	Quantitative Composition (%, w/w)														
	FA	FA2	FB	FB2	FC	FC2	FD	FD2	FD3	FE	FE2	FF	FF2	FF3	FH
CMC	1.5	2													
MC			1.5	2											
HEC					1.5	2									
XG							2.5	1.5	1						
Gelatine										5	1.5	10	2	3	
HA															1.5
S-Methylparaben	0.18	0.18	0.18	0.18	0.18	0.18	0.18	0.18	0.18	0.18	0.18	0.18	0.18	0.18	0.18
S-Propylparaben	0.02	0.02	0.02	0.02	0.02	0.02	0.02	0.02	0.02	0.02	0.02	0.02	0.02	0.02	0.02
Sucralose	0.5	0.5	0.5	0.5	0.5	0.5	0.5	0.5	0.5	0.5	0.5	0.5	0.5	0.5	0.5
Citric Acid	0.05	0.05	0.05	0.05	0.05	0.05	0.05	0.05	0.05	0.05	0.05	0.05	0.05	0.05	0.05
Sodium Citrate	0.5	0.5	0.5	0.5	0.5	0.5	0.5	0.5	0.5	0.5	0.5	0.5	0.5	0.5	0.5
Purified Water	qs 100	qs 100	qs 100	qs 100	qs 100	qs 100	qs 100	qs 100	qs 100	qs 100	qs 100	qs 100	qs 100	qs 100	qs 100

3.2 Rheological Characterization

The rheological characterization of the formulations was performed by observing the behaviour of viscosity as a function of shear rate. In Fig. 1 it is possible to observe that the two hydrogels more similar to Gloup® are FA2 and FD3. Their viscosity decreases as the shear rate applied increases, so they behave like shear thinning fluids.

Based on these results, and on macroscopic evaluation, Formulations FE and FF were eliminated from the study.

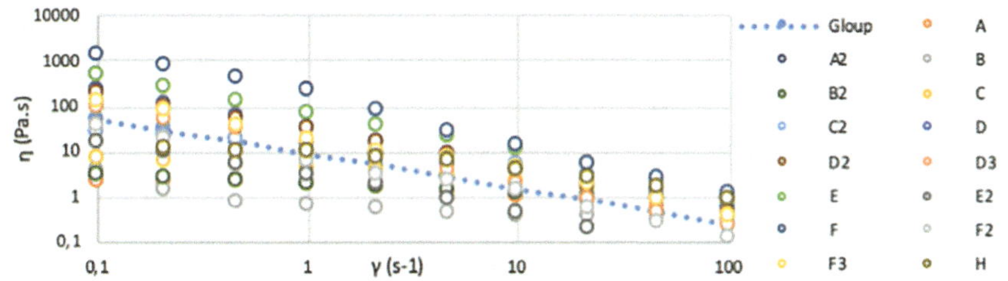

Fig. 1. Apparent viscosity vs shear rate for all formulations.

Concerning the oscillation frequency test, all formulations were tested, but we will focus mainly on the results of formulations FA2 and FD3. In Fig. 2, both Gloup® and Formulation FD3 exhibited predominantly elastic behaviour, evident from the greater magnitude of the elastic module (G') to that of the viscous module (G"). This means that the structure of the gel remained intact through the entire range of frequencies, confirming that the two hydrogels present a strong network and present a solid-like behaviour (G' > G").

For Formulation FA2, at lower frequencies the G" is superior, but at the highest frequencies a transition can be observed, where the G' is found to be higher that G". Studies show that this type of behaviour is commonly seen in formulations using concentrations of CMC close to 2.5% and could be related to the formation of a three-dimensional network [2].

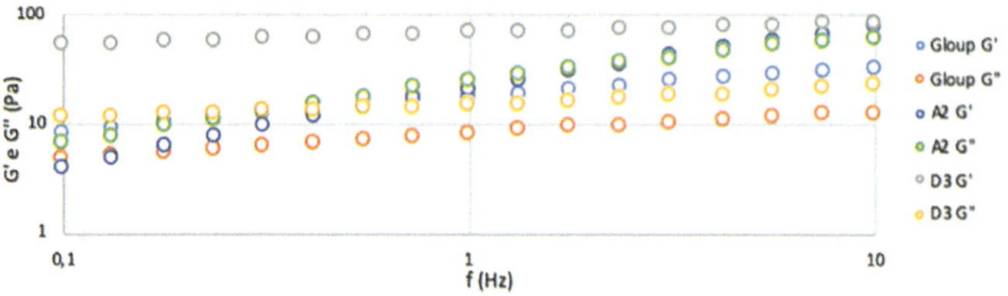

Fig. 2. Frequency sweep test for Gloup® and formulations FA2 and FD3.

To measure the mucoadhesion, a rheological method was performed in a simulated physiological environment using mucin and artificial saliva in the proportion 1:1. The behaviour of polymer candidates was determined by comparing the rheological properties of polymer-mucin and polymer-artificial saliva mixtures with those of the polymers separately. All vehicles exhibited a higher viscosity in the mixtures (Fig. 3), showing rheological synergism. Between the two final formulations, the one whose results were more affected by the presence of mucin, or artificial saliva, was Formulation FA2. This means that CMC is likely to form more adhesive bonds with the mucus at the time of oral intake than XG.

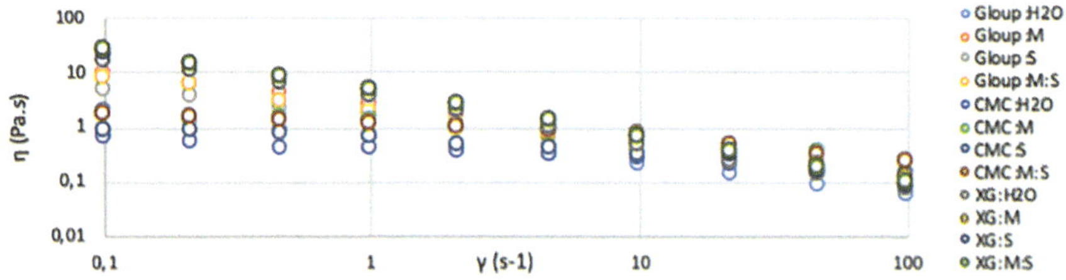

Fig. 3. Influence of mucin and artificial saliva: apparent viscosity vs shear rate for Gloup® and formulations FA2 and FD3

The same could be seen in the results of the oscillation test (data not shown), once the elastic module showed superior values in the mixtures comparing to the polymers alone, this translates as a positive interaction caused by mucoadhesion.

Regarding the results of the adhesion test (data not shown), XG showed more adhesive proprieties because it has higher values of the area under force time. However, if we compare the results of the polymer alone and in the mixtures, they are all very similar. CMC, despite showing less adhesive strength, is more affected by the presence of mucin and artificial saliva. Comparing the two hydrogels with Gloup®, we can state that the one that comes closer to its mucoadhesive behaviour is CMC. Both formulations showed a greater ability to form bonds with the mucin and the artificial saliva, which means that they will remain longer in the oral cavity at the time of administration. On the other hand, XG was not as affected by the mixtures and this could be an advantage for the oral intake of the vehicle.

4 Conclusions

From all the initial formulations, only two were selected as final formulations based on several rheological characterization tests. The chosen polymers were CMC at 2% and XG at 1%. Further testing shall be done in order to evaluate their stability over time and to predict interactions between the vehicle and the active substances/medicines at the time of administration.

References

1. Ivanovska, V., Rademaker, C.M.A., van Dijk, L., Mantel-Teeuwisse, A.K.: Pediatrics **134**(2), 361–372 (2014)
2. Benchabane, A., Bekkour, K.: Colloid Polymer Sci. **286**, 1173 (2008)

Choosing Sustainable Alternatives for Cosmetic Emollients: Sustainability vs Rheological Performance

Sara Bom, Helena Margarida Ribeiro, and Joana Marto(✉)

Research Institute for Medicines (iMed.ULisboa), Faculty of Pharmacy, Universidade de Lisboa, Lisbon, Portugal
{sarabom,hribeiro}@campus.ul.pt, jmmarto@ff.ulisboa.pt

Abstract. Understanding the physicochemical properties of sustainable raw materials plays a crucial role in the formulation process, since it allows to predict formulation issues normally interlinked with this type of compounds. The main goal of this work was to design specific tools to forecast the spreadability of sustainable raw materials. To meet this goal, 5 alternatives for petrolatum (P) and 5 alternatives for dimethicone (D) were selected. The experiments included *in vitro* and *in vivo* spreading behavior characterization and viscosity measurements. For P alternatives, the spreading value was best fitted as a function of the logarithm of viscosity and for D alternatives the spreading value was best fitted as a direct function of viscosity. The mathematical model defined for P alternatives could be used for the most diverse chemical and physical types of alternatives, which include butter-like, jelly-like and wax-like raw materials; the mathematical model defined for D could be used for high, medium and low viscosity raw materials. In conclusion, the application of these tools will help the entire replacement process and will also contribute to the development of sustainable cosmetic products with higher performance.

1 Introduction

Cosmetic industry is trying to embrace the sustainability concept in their industrial practices and market products. However, there is still a long away to go in this area, mainly to the level of choice of raw materials that are in accordance with the concept of sustainability. When choosing a sustainable cosmetic product, the consumer expects the quality and efficacy of a "non-sustainable" alternative and this has been one of the great challenges of the cosmetics industry [1]. Ingredients that meet the sustainable concept have been associated with formulation issues, regarding its performance, aesthetics, sensorial properties and marketability. Petrolatum (P) and dimethicone (D) are among the most used emollient ingredients in the cosmetic industry due to their function and properties within a formulation and their replacement can be a rather difficult task, being extremely necessary to understand beforehand the behavior of the possible alternative ingredients [1, 2]. Thus, understanding the correlation between the chemical and physicochemical properties of these compounds then plays a crucial role

F. J. Galindo-Rosales et al. (Eds.): IBEREO 2019, SPM, pp. 40–44, 2020.
https://doi.org/10.1007/978-3-030-27701-7_9

in the formulation process, since it allows predicting stability, performance and aesthetic problems [2]. Spreadability appears in this context because is one of the most valuable physicochemical properties according to the consumer [3].

2 Materials and Methods

2.1 Materials

Petrolatum and dimethicone alternatives were selected taking in account its molecular structure (i.e. chemical properties), physicochemical properties and sustainability criteria. Petrolatum alternatives include: NVA (Ricinus Communis Seed Oil; Hydrogenated Castor Oil; Copernicia Cerifera Cera), SB (Butyrospermum Parkii), PB (Butyrospermum Parkii (Shea) Butter, Elaeis Guineensis (Palm) Butter, Simmondsia Chinensis (Jojoba) Seed Oil, Punica Granatum Fruit Juice), MB (Mangifera Indica (Mango) Seed Butter), and KV (Ricinus Communis (Castor) Seed Oil (and) Hydrogenated Rhus Verniciflua Peel Wax (and) Rhus Succedanea Fruit Wax (and) Ascorbyl Palmitate (and) Tocopherol). D alternatives include: PLS (Hydrogenated Ethylhexyl Olivate (and) Hydrogenated Olive Oil Unsaponifiables), EMG (C15-19 Alkane), MOD (Octyldodecyl Myristate), DPPG (Propylene Glycol Dipelargonate), and, SQ (Hydrogenated Polyisobutene).

2.2 Methods

2.2.1 Spreading Behavior Characterization

For spreading behavior characterization 3 methods were applied: spreading value determination - area (mm^2), contact angle measurements (mean ± SD), and, *in vivo* spreading measurements (14 volunteers, n = 2).

2.2.2 Viscosity Measurements

Viscosity measurements were performed using a controlled stress Kinexus Rheometer (Malvern) using a plate-plate geometry for P alternatives and a bob-in-cup geometry for D alternatives. The measurements were carried out between 1 and 1000 Pa on a logarithmic increment, ranging from 0.1 to 100 s^{-1}, at 25 °C.

2.2.3 Tools Development

The spreading value was defined as the values obtained through the spreading value determination method – area (mm^2). The viscosity value was defined as the value corresponding to a shear rate of 46.42 s^{-1} (for P alternatives) and to a shear rate of 21.55 s^{-1} (for D alternatives), reported from the curve representing the viscosity as a function of shear rate. Linear regressions were used to express spreading value as a function of viscosity.

3 Results and Discussion

3.1 Spreading Behavior Characterization

Firstly, the spreading value determination - area (mm^2) and contact angle measurements were used to evaluate the spreading properties of sustainable emollients, in order to evaluate with which method is possible to achieve the best correlation of results. To note that these tests were performed in an "artificial" substrate, Sil-Tec, that pretends to mimic the natural barrier of healthy skin due to the porous nature of the material. Spreading value determination and contact angle measurements are often used to evaluate the spreading behavior of liquids, but with the results obtained for P alternatives (Table 1) is possible to affirm that is indeed possible to apply these methods to evaluate solid-like raw materials. During the test it was essential to control the temperature to obtain reproducible rheological characteristics due to the fact that this type of emollients are temperature sensitive materials; this is consistent with the results obtained by Pandey *et al.* [4]. In general terms, through a meticulous comparison between the two first methods, the spreading value provided better discrimination of the selected emollients and the results obtained have a better correlation with the viscosity data obtained. In terms of the data provided by the *in vivo* spreading method, it can be stated that the consumer opinion correlates positively with the data obtained with the other two methodologies, except for the results for sample SB. Considering all these features, only the spreading value (i.e. area) was considered in the following studies.

Table 1. Synthesis of the results obtained in terms of spreadability.

Petrolatum alternatives			Dimethicone alternatives		
Identification name	Area (mm^2)	Contact angle (°)*	Identification name	Area (mm^2)	Contact angle (°)*
P	12.6	100.40 ± 1.07	D	188.5	24.23 ± 0.42
NVA	28.3	71.10 ± 0.36	PLS	102.1	56.13 ± 1.40
SB	19.6	102.23 ± 0.81	EMG	117.8	55.83 ± 0.23
PB	15.7	97.13 ± 1.60	MOD	102.1	53.87 ± 1.21
MB	38.5	59.33 ± 1.93	DPPG	127.2	53.33 ± 0.75
KV	50.3	38.40 ± 0.57	SQ	70.7	61.73 ± 2.08

*(mean ± SD, n = 3)

3.2 Tools Development

As stated in the experimental section, linear regressions were used and for each case was analyzed which mathematical model bests fitted the spreading value as a function of viscosity. For P alternatives, the spreading value was best fitted as a function of logarithm of viscosity. For D alternatives, the spreading value was best fitted as a direct function of viscosity (Fig. 1).

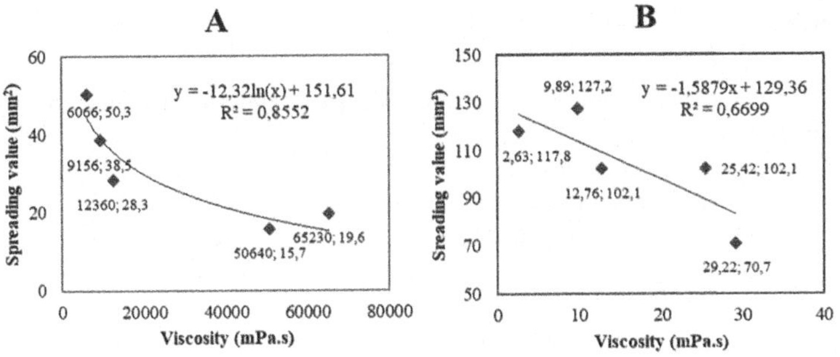

Fig. 1. Adjustment model for spreading value versus viscosity – P alternatives (A) and D alternatives (B).

The value of intercept b depends on the emollient family and the slope "a" of the model equations are negative whatever the emollient. This conclusion is in agreement with the results obtained in the work performed by Douguet *et al.* [3]. The different patterns revealed by the mathematical models allows to indicate that the viscosity of the emollients strongly influences the spreading behavior which will have a large influence on the performance of cosmetic formulations.

3.3 Proof of Concept

The proof of concept was designed in order to evaluate the maximum number of alternatives and to verify the spreading value of each alternative selected for the both emollient groups. In specific, for P alternatives it was decided to test three more butter-like ingredients and one wax-like raw material. In relation to the application range of the model mathematical equation defined for the alternatives to D, since high, medium and low viscosity raw materials were included for the determination of the tools there was no need to evaluate more ingredients to prove the veracity and application range of the results extracted. Mathematical model defined for P alternatives could be used for the most diverse chemical and physical types of alternatives, which include butter-like, jelly-like and wax-like raw materials; the mathematical model defined for D could be used for high, medium and low viscosity raw materials.

4 Conclusions

This work describes the successful development and assessment of mathematical models intended to predict the emollient property of several sustainable ingredients, which can be considered of extreme importance given the urgent need to resort to effective and quickly reformulation strategies. This tool allows the formulator to evaluate one of the most important characteristics of the formulations from the point of view of the consumer, the spreadability. Additionally, since the results obtained with the *in vivo* spreading characterization method shows a positive correlation with the spreading determination data utilized, it is possible to affirm that these tools are in-line with the consumer opinion.

References

1. Sahota, A. (ed.): Sustainability: How the Cosmetics Industry Is Greening Up. Wiley, West Sussex (2014)
2. Bom, S., Jorge, J., Ribeiro, H.M., Marto, J.: A step forward on sustainability in the cosmetics industry: a review. J. Clean. Prod. **225**, 270–290 (2019)
3. Douguet, M., Picard, C., Savary, G., Merlaud, F., Loubat-bouleuc, N., Grisel, M.: Spreading properties of cosmetic emollients: use of synthetic skin surface to elucidate structural effect. Colloids Surf. B Biointerfaces **154**, 307–314 (2017)
4. Pandey, P., Ewing, G.D.: Rheological characterization of petrolatum using a controlled stress rheometer. Drug Dev. Ind. Pharm. **34**(2), 157–163 (2008)

Development of an Emulgel Type Format Based on Rosemary Essential Oil and a Fumed Silica. Influence of a Shear Post-treatment on Its Final Properties

Jenifer Santos García[1](✉), María José Martín-Piñero[2], José Antonio Carmona[2], Nuria Calero[2], and José Muñoz[2]

[1] Departamento de Ingeniería Química, Escuela Politecnica Superior, Universidad de Sevilla, C/Virgen de África, 7, 41007 Seville, Spain
jsantosgarcia@us.es

[2] Departamento de Ingeniería Química, Facultad de Química, Universidad de Sevilla, C/Profesor García Gonzalez, 1, 41012 Seville, Spain
{mjmartin, joseacarmona, nuriacalero, jmunoz}@us.es

Abstract. The main objective of this work was to study the influence of the gelling agent (Aerosil 200) concentration on viscoelastic functions. A further objective was to study the effect of different shear-induced post-treatments on the viscoelastic functions and microstructure of rosemary essential oil/Aerosil 200 emulgels. The increase of Aerosil 200 concentration provoked an increase of viscoelastic functions. In addition, the shear applied influences both elastic and viscous modulus. Interestingly, there is a decrease in viscoelastic properties with shear. This work proves the importance of controlling shear post-treatments.

1 Introduction

Oil-in-water emulsions (oil droplets dispersed in an aqueous matrix) present many applications in different fields such as cosmetics, food, pharmaceutics and agrochemicals. In addition, gel emulsions (also called emulgels) are very interesting format for the development of cosmetic products. This format exhibits the advantages of emulsions and gels to encapsulate active ingredients for topical and transdermal use.

Essential oils are derived directly from natural sources and they are attracting much attention due to the interest of consumers in using natural products. Rosemary essential oil, extracted from *Rosmarinus officinalis Lamiaceae,* has been used in different applications such as flavouring agent and food preservative [1]. This fact is due to its antioxidant and antibacterial activities [2]. However, essential oils show a high volatility, which is an important drawback that can be overcome by using an emulgel format as a way of encapsulating active ingredients.

Emulgels are composed of an emulsion and a gelling agent. Different compounds can be added to the emulsion to form the emulgel, e.g. polysaccharides, clays and oxide gels. Aerosil 200, a fumed silica, can be used as gelling agent since it forms a 3D network in aqueous solutions [3].

F. J. Galindo-Rosales et al. (Eds.): IBEREO 2019, SPM, pp. 45–48, 2020.
https://doi.org/10.1007/978-3-030-27701-7_10

The microstructure of emulsions could be modified by shear during mixing, processing and transport operations. In addition, shear-based processes may alter the functional properties of dispersed systems such as its viscosity, viscoelastic functions or its physical stability. Hence, it is important to tightly control the preparation of emulgels.

2 Materials and Methods

2.1 Materials

The dispersed phase used was rosemary essential oil (Sigma Chemical Company). The continuous phase was composed by Aerosil 200 particles (Quimidroga, EVONIK), and Pluronic PE9400 (BASF).

2.2 Emulgels Developed

Aerosil suspensions were prepared by using a magnetic stirrer for 3 h, whose pH were adjusted to 7. 2 wt% of Pluronic PE9400 was dispersed into miliQ water. The essential oil used (20 wt%) was homogenized using a Silverson L5M at 4000 rpm for 45 s. Then, this coarse emulsion was passed twice through Microfluidizer M110P at 5000 psi. Finally, the corresponding Aerosil suspension was incorporated into the microfluidized emulsion in a ratio of 1:1 using a magnetic stirrer for 10 min.

2.3 Rheology

Rheological measurements were carried out by using a controlled-stress Haake MARS II rheometer. In order to study the influence of shear on the viscoelastic functions of the emulgels studied, a frequency sweep test was carried out from 20 rad/s to 0.05 rad/s before and after the shear post-treatment. These shear post-treatments consisted of start-up at the inception of shear experiments, which were conducted for 600 s at five different shear rates (25, 50, 100, 250 and 500 s^{-1}).

2.4 Cryo-Scanning Electron Microscopy

The microstructure of the emulgels before and after the shear post-treatments was observed using a Cryo Scanning Electron microscope (Zeiss EVO) at 8 kV. The sample preparation method was the same reported by Santos et al. [4].

3 Results and Discussion

Figure 1 shows the elastic (G’) and viscous (G”) moduli at 1 rad/s as a function of Aerosil 200 concentration for the emulgels developed. The elastic modulus is higher than the viscous modulus in all the cases studied. This fact is a clear evidence of the formation of a gel, thanks to the incorporation of Aerosil 200. Interestingly, the emulsion prepared without Aerosil 200 did not show measurable viscoelastic properties (data not shown). In addition, there is an increase of both viscoelastic functions with Aerosil concentration. Nevertheless, G' and G" for 5 wt% and 7 wt% emulgels were not significantly different. This indicates the formation of a stronger gel at the highest Aerosil concentration. Aerosil 200 is able to protect the interface and stabilize the emulgels formed. The increase in viscoelastic functions from 7 to 8 wt% Aerosil could be due to the excess of Aerosil that can remain in the continuous phase.

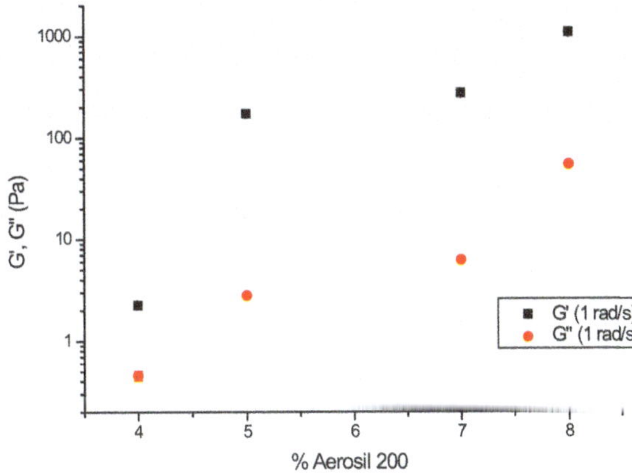

Fig. 1. Elastic and viscous moduli at 1 rad/s as a function of Aerosil concentration. T = 20 °C.

Table 1. Influence of different shear post-treatments on the viscoelastic functions for 8 wt% emulgel. T = 20 °C.

Shear rate applied (s^{-1})	G' (Pa)	G" (Pa)
0	1087	56
25	591	62
50	472	53
100	461	48
250	384	33
500	380	32

The emulgel containing 8 wt% of Aerosil 200 was selected to study different shear post-treatments. Table 1 shows G' and G" before (0 s^{-1}) and after the shear post-treatment applied for 5 min. G' markedly decreased with the post shear rate used. In addition, G' had a tendency to level off after 100 s^{-1}. These facts indicate the partial break-up of the gel structure formed with Aerosil 200 with the shear post-treatment.

Figure 2A and B show the microstructure of 8 wt% emulgel before and after a shear post-treatment of 500 s^{-1}, respectively. The microphotograph of the emulgel without shear post-treatment shows a more aggregated structure. This structure is caused by Aerosil particles. Once the shear post-treatment is applied and finished (Fig. 2B), the structure seems to be oriented in the direction of the flow and less aggregated. Hence, the shear post-treatment could break some aggregates, which may be positive for the physical stability of these emulgels.

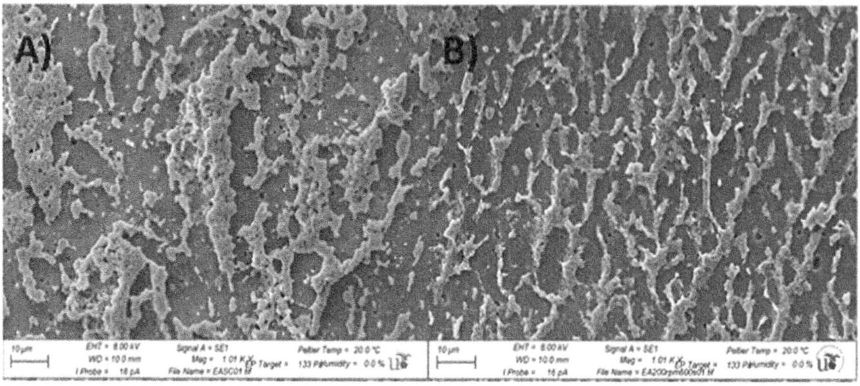

Fig. 2. (A and B). Microstructure observed by Cryo-SEM of 8 wt% emulgel without any shear post-treatment and after a shear post-treatment of 500 s^{-1} for 600 s.

4 Conclusions

Emulgels formulated with Aerosil 200 and rosemary essential oil were developed. Stronger gels were obtained using higher Aerosil 200 concentration. Shear post-treatments provoked the decrease in the viscoelastic functions due to the break of some aggregates. Therefore, a tight control of shear post-treatments, such as transport, performance or mixing, is recommended.

Acknowledgements. The financial support received (Project CTQ2015-70700-P) from the Spanish Ministerio de Economia y Competitividad and the European Commission (FEDER Programme) is kindly acknowledged.

References

1. Couto, R.O., Conceição, E.C., Chaul, L.T., Oliveira, E.M., Martins, F.S., Bara, M.T.F., Paula, J.R.: Spray-dried rosemary extracts: physicochemical and antioxidant properties. Food Chem. **131**, 99–105 (2012)
2. Baser, K.H.C., Buchbauer, G.: Handbook of Essential Oils: Science, Technology, and Applications. CRC Press, Boca Raton (2015)
3. Heath, D., Tadros, T.F.: Influence of pH, electrolyte, and poly (vinyl alcohol) addition on the rheological behavior of aqueous silica (Aerosil) dispersions. J. Colloid Interface Sci. **93**, 320–328 (1983)
4. Santos, J., Calero, N., Trujillo-Cayado, L., Alfaro, M.C., Muñoz, J.: The role of processing temperature in flocculated emulsions. Ind. Eng. Chem. Res. **57**, 807–812 (2017)

Effect of Chemical Composition on the Thermal Profiles of Afuega'l Pitu Cheese (PDO)

Lorena Piñeiro[1(✉)], Inmaculada Franco[1], and Clara A. Tovar[2]

[1] Food Technology Area, Faculty of Sciences, University of Vigo, Campus Universitario As Lagoas s/n, 32004 Ourense, Spain
{lopineiro, inmatec}@uvigo.es

[2] Department of Applied Physics, Faculty of Sciences, University of Vigo, Campus Universitario As Lagoas s/n, 32004 Ourense, Spain
tovar@uvigo.es

Abstract. The objective of this study was to know the thermo-rheological behaviour of eight batches of Afuega'l Pitu cheese (F1−F8), 30 days ripened, based on its biochemical composition. Negative correlations between the moisture-to-protein ratio and the viscoelastic moduli (G' and G'') from dynamic thermo-mechanical analysis (20 °C−90 °C) were determined. F1–F8 had similar thermal profiles maintaining the solid-like character in the protein matrices up to 90 °C.

1 Introduction

Afuega'l Pitu is an artisanal cheese made in Asturias (northern Spain) from whole cow's milk by lactic coagulation. It has been protected with a Designation of Origin (PDO) since 2004 [1]. Four varieties can be distinguished: with truncated cone (*atroncau*) or courgette (*trapu*) shape, and whether paprika is added (*roxu*) or not (*blancu*). The elaboration is slow and unhurried, resulting in a soft paste which may be consumed either fresh or ripened. Acid milk gels are formed at pH near the isoelectric point of the caseins by a massive colloidal aggregation and gelation into the final curd [2]. The properties of the network may be modified by the protein content, as well as by the fat and moisture state and content, which influence the viscoelastic properties of cheese. Nevertheless, the temperature might alter the gel structure due to the melting of cheese fat or enhancing hydrophobic interactions in the protein network [3]. To date, there were no data regarding thermal profiles in terms of the viscoelastic moduli (G' and G'') and loss factor ($tan\delta$) of Afuega'l Pitu cheese. The objective of this work was to determine the thermo-rheological response of Afuega'l Pitu cheese *atroncau blancu* variety. These data will be discussed based on the biochemical characteristics of the different samples.

F. J. Galindo-Rosales et al. (Eds.): IBEREO 2019, SPM, pp. 49–53, 2020.
https://doi.org/10.1007/978-3-030-27701-7_11

2 Materials and Methods

2.1 Cheesemaking and Sampling

Sixteen cheeses (ripened for 30 days) were manufactured in eight farms (F1–F8) following the PDO specifications. Samples were divided into two halves. One half for biochemical analysis was triturated without rind and hermetically stored at −30 °C, and the other was used in the rheological tests on the same day.

2.2 Biochemical Determinations

Total solids (TS), fat, protein, pH and titratable acidity (TA) were analysed following [4], ash and fat acidity (FA) by IDF standards [5, 6], NaCl by the AOAC method [7]. Water activity (a_w) was measured with LabMaster-aw instrument (Novasina AG, Switzerland); Ca, K, and Na by atomic absorption with an air/acetylene flame, and P by inductively coupled plasma optical emission spectroscopy (ICP-OES) with a Varian SpectrAA 220 FS absorption spectrophotometer (Varian, Inc., USA).

2.3 Thermo-Rheological Analysis

Samples were allowed to stand for 15 min to achieve thermal and mechanical equilibrium before measurement with a Haake RS600 stress-controlled rheometer (Thermo Electron Corp., Germany) using a PP20 (1 mm gap) in a solvent trap. Temperature was controlled using a Peltier system (± 0.1 °C). Dynamic thermo-mechanical analysis (DTMA) were performed from 20 °C to 90 °C at 1 °C/min, at 0.1 Hz frequency and strain ($\gamma = 0.1\%$), within the LVE range [8].

2.4 Statistical Analysis

ANOVA using Tukey HSD test (biochemical data) and Pearson correlation coefficient ($^{***}P < 0.001$, $^{**}P < 0.01$, $^{*}P < 0.05$) were performed with the Statistical 8.0 program for Windows (Statsoft Inc., USA). Rheological data were expressed as mean values of at least three independent measurements.

3 Results and Discussion

3.1 Biochemical Parameters

Biochemical parameters showed slight but significant differences among of F1–F8 (Table 1). Regarding physico-chemical parameters, F1–F8 presented pHs near to the casein isoelectric-point (pI ~ 4.6) due to the mainly acid coagulation. In addition, practically all the water present in F1–F8 is available to interact with solutes and surfaces (a_w ~ 0.99). The TA indicates the total acidity of the cheese, exhibiting F2 and F6 the highest values. F8 exhibited the highest FA indicating a greater degree of fat hydrolysis.

Table 1. Afuega'l Pitu cheese *atroncau blancu* variety samples (F1−F8) biochemical parameters (TS, total solids; S/M, salt in moisture; FPR, fat-to-protein ratio; MPR, moisture-to-protein ratio; TA, titratable acidity; FA, fat acidity)

	TS (g/100 g)	Ash (g/100 g TS)	Fat (g/100 g TS)	Protein (g/100 g TS)	S/M (g NaCl/100 g moisture)	NaCl (g/100 g TS)	FPR	MPR
F1	55.2 ± 0.6[a]	2.5 ± 0.2[a]	53 ± 1[a]	37.1 ± 0.4[ac]	2.1 ± 0.2[a]	1.7 ± 0.2[a]	1.44 ± 0.06[a]	2.2 ± 0.1[a]
F2	58 ± 3[ab]	3.9 ± 0.7[b]	50 ± 1[b]	39 ± 1[b]	3.3 ± 0.2[bc]	2.4 ± 0.2[b]	1.30 ± 0.07[b]	1.9 ± 0.2[b]
F3	62 ± 1[b]	2.6 ± 0.1[ac]	51 ± 1[ab]	36.0 ± 0.3[acd]	2.6 ± 0.3[ab]	1.6 ± 0.1[a]	1.42 ± 0.05[ab]	1.7 ± 0.1[b]
F4	61.8 ± 0.3[b]	3.9 ± 0.3[b]	52 ± 2[ab]	36 ± 2[ad]	4.0 ± 0.2 [cd]	2.5 ± 0.1[b]	1.46 ± 0.13[ab]	1.7 ± 0.1[b]
F5	66 ± 2[c]	3.35 ± 0.02[bc]	50.5 ± 0.7[b]	38.0 ± 0.5[bc]	4.6 ± 0.3[d]	2.36 ± 0.08[b]	1.33 ± 0.04[b]	1.4 ± 0.1[c]
F6	59 ± 2[b]	3.5 ± 0.3[b]	51.5 ± 0.5[ab]	35.7 ± 0.8[ad]	4.2 ± 0.3[d]	2.8 ± 0.1[b]	1.44 ± 0.05[a]	1.9 ± 0.1[b]
F7	51 ± 2[d]	3.3 ± 0.4[abc]	53 ± 1[ab]	34.6 ± 0.7[d]	2.5 ± 0.7[a]	2.4 ± 0.5[b]	1.53 ± 0.06[a]	2.8 ± 0.01[d]
F8	48.6 ± 0.1[d]	4.0 ± 0.1[b]	51.6 ± 0.8[ab]	36.1 ± 0.6[acd]	2.5 ± 0.2[a]	2.6 ± 0.2[b]	1.43 ± 0.04[a]	2.9 ± 0.1[d]
	Na (g/kg TS)	K (g/kg TS)	Ca (g/kg TS)	P (g/kg TS)	pH	TA (g lactic ac./100 g TS)	a_w	FA (mg KOH/g fat)
F1	4 ± 1[a]	3.61 ± 0.07[ab]	2.45 ± 0.06[a]	5.7 ± 0.3[a]	4.32 ± 0.04[af]	1.30 ± 0.06[a]	0.998 ± 0.003[ab]	1.1 ± 0.3[ac]
F2	10 ± 4[a]	3.71 ± 0.09[ab]	2.2 ± 0.3[a]	5.65 ± 0.04[a]	4.27 ± 0.02[ac]	1.55 ± 0.05[b]	0.995 ± 0.001[a]	0.85 ± 0.06[ae]
F3	4.1 ± 0.7[a]	3.74 ± 0.03[ab]	2.06 ± 0.09[a]	5.7 ± 0.2[a]	4.38 ± 0.03[b]	1.37 ± 0.02[a]	0.999 ± 0.001[ab]	0.39 ± 0.04[b]
F4	10 ± 2[a]	3.4 ± 0.3[a]	2.0 ± 0.3[ab]	5.8 ± 0.4[a]	4.34 ± 0.02[bf]	1.34 ± 0.04[a]	0.998 ± 0.002[ab]	0.65 ± 0.02[abe]
F5	8.0 ± 0.2[a]	3.64 ± 0.01[ab]	1.9 ± 0.1[ab]	5.3 ± 0.2[ab]	4.22 ± 0.01[c]	1.32 ± 0.04[a]	0.999 ± 0.002[ab]	1.4 ± 0.1[c]
F6	9 ± 1[a]	3.94 ± 0.08[b]	2.09 ± 0.02[a]	5.7 ± 0.3[a]	4.09 ± 0.02[d]	1.65 ± 0.03[b]	1.000 ± 0.001[b]	0.57 ± 0.06[be]
F7	8 ± 2[a]	3.49 ± 0.06[ab]	2.10 ± 0.01[a]	5.33 ± 0.00[ab]	4.14 ± 0.02[d]	1.34 ± 0.05[a]	1.000 ± 0.000[b]	0.29 ± 0.04[b]
F8	11.5 ± 0.2[a]	4.97 ± 0.00[c]	1.37 ± 0.01[b]	4.71 ± 0.01[b]	4.52 ± 0.02[e]	1.26 ± 0.08[a]	1.000 ± 0.001[b]	2.7 ± 0.3[d]

[a−e]indicates significant differences ($P < 0.05$).

Regarding biochemical parameters, F5 showed the highest TS content and the lowest moisture-to-protein ratio (MPR), whereas F7 and F8 exhibited the lowest TS content and the highest MPR. F1 had higher fat content than F2 and F5, whose fat contents and fat-to-protein ratios (FPR) were the lowest. As to NaCl content, F1 and F3 showed the lowest values, in line with a lower ash and Na content. Nevertheless, F4 and F5 (high TS), and F6 (high NaCl content) exhibited significantly higher salt in moisture (S/M) content. Na and P were the most abundant minerals in F1–F8, while Ca content was the lowest. Since Afuega'l Pitu was acid-coagulated [2], a significant demineralisation of casein micelles occurs compared to enzymatically coagulated cheeses [9].

3.2 Thermo-Rheological Properties

F1−F8 thermal profiles showed higher storage (G') than loss (G'') moduli from 20 °C to 90 °C, indicating the solid-like behaviour permanence during heating (Fig. 1a, b). This trend can be explained based on the low pH (~pI), as the neutral electrostatic charge enhances the attractive forces between caseins reinforcing the compaction strength in the casein matrix.

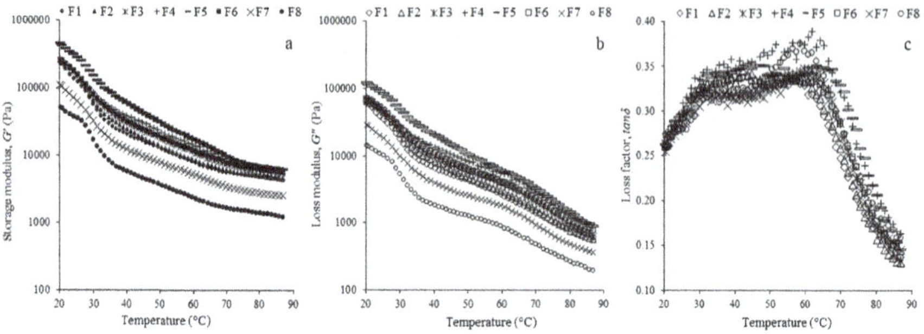

Fig. 1. Thermal profiles of Afuega'l Pitu cheese, storage modulus, G' (a), loss modulus, G'' (b), and loss factor, *tan δ* (c) at 0.1 Hz

F5 showed the highest G' and G'' throughout the temperature range (Fig. 1a, b), which could be related to the lowest MPR. Conversely, F7 and F8 exhibited the lowest G' and G'' showing the greatest MPR. These results were in line with the Pearson correlations between G'–MPR (-0.87^{***}) and G''–MPR (-0.88^{***}).

From 20 °C to 60 °C, G' and G'' of F1–F8 showed a steadily decrease, being the thermal decrease of G' more intense than that of G'', increasing $tan\delta$ in this specific temperature range (Fig. 1). These thermal responses indicate a network weakening associated with a loss of the solid-like character, due to the disruption of the polar interactions and hydrogen bonds in the micellar structure [10]. Moreover, fat globules melting (~40 °C) may also contribute to the continued softening of the structure [2]. Then, G' and G'' continued to decrease up to 90 °C showing a net softening of F1–F8 networks. Nevertheless, G' decreased more slowly than G'' with temperature, and hence $tan\delta$ decreased notably showing an increase in the solid-like nature of F1–F8 gel networks (Fig. 1). That means a longer lifetime in the casein-casein interactions indicating the presence of more energy stable bonds in a more ordered casein matrix [11]. Moreover, heating at high temperature while shearing results in casein dehydration and free fat formation, so, casein matrix constricts, moisture leaks, and fat exudates [2], promoting new hydrophobic interactions among caseins. This effect could be reinforced by shearing [8].

4 Conclusions

Regardless of the differences found in the chemical composition between F1–F8, thermal profiles were similar for all cheeses. The lowest viscoelastic moduli, corresponding to the largest hydrated networks, were maintained throughout the heating. At high temperatures, all samples showed an increase in the solid-like character together with a softening of the cheese structure.

References

1. Commission Regulation (EC) No 723/2008. OJ L 198 (2008)
2. Guinee, T.P.: Role of protein in cheese and cheese products. In: Fox, P.F., McSweeney, P.L.H. (eds.) Advanced Dairy Chemistry—1 Proteins, pp. 1083–1174. Springer, Boston (2003)
3. Norton, I.T., Spyropoulos, F., Cox, P.: Practical Food Rheology. Blackwell Publishing Ltd., Hoboken (2011)
4. Bargiela, V., Franco, I., Tovar, C.A.: Queso San Simón da Costa (DOP). Editorial Académica Española, Saarbrücken (2012)
5. International Dairy Federation, IDF Standard No. 27. (IDF, Brussels) (1964)
6. International Dairy Federation, IDF Standard No. 6. (IDF, Brussels) (2004)
7. Horwitz, W.: Official Methods of Analysis of the Association of Official Analytical Chemists, 18th edn. AOAC Intel, Gaithersburg (2005)
8. Piñeiro, L., Franco, I., Tovar, C.A.: Rheological and biochemical study of Afuega'l Pitu cheese (PDO). In: Hernández, M.J., Sanz, T., Salvador, A., Rubio-Hernández, F.J., Steinbrüggen, R. (eds.) The Multidisciplinary Science of Rheology. Towards a Healthy and Sustainable Development, pp. 56–59 (2017)
9. Zamberlin, S., Antunac, N., Havranek, J., Samarzija, D., Mljekarstvo, D.: Mineral elements in milk and dairy products. Mljekarstvo **62**, 111–125 (2012)
10. Damodaran, S., Paraf, A.: Food Proteins and Their Applications. Marcel Dekker, New York (1997)
11. Renkema, J.M.S., van Vliet, T.: Heat-induced gel formation by soy proteins at neutral pH. Agric. Food Chem. **50**, 1569–1573 (2002)

Microstructure, Rheology, and Composition of a Spanish Cheese

Lorena Piñeiro[1(✉)], Inmaculada Franco[1], Clara A. Tovar[2], and Laura Campo-Deaño[3]

[1] Food Technology Area, Faculty of Sciences, University of Vigo, Campus Universitario As Lagoas s/n, 32004 Ourense, Spain
{lopineiro,inmatec}@uvigo.es

[2] Department of Applied Physics, Faculty of Sciences, University of Vigo, Campus Universitario As Lagoas s/n, 32004 Ourense, Spain
tovar@uvigo.es

[3] CEFT, Departamento de Engenharia Mecânica, Faculdade de Engenharia da Universidade do Porto, Rua Dr. Roberto Frias, 4200-465 Porto, Portugal
campo@fe.up.pt

Abstract. Afuega'l Pitu *atroncau* cheese, *Blancu* (B) and *Roxu* (R) varieties, were collected at 30 days of ripening from 9 farms (B1−B9 and R1−R9) to determine the biochemical and linear viscoelastic parameters, as well as the microstructure. B and R samples exhibited fused casein particles in the protein matrix with a high packing effect and different fat distribution. The lowest total solids content led to the lowest stress amplitude (σ_{max}) and gel strength (G^*). B samples exhibited higher values of strain amplitude (γ_{max}) and σ_{max} than R, but with a similar solid-like character ($tan\delta$).

1 Introduction

In Afuega'l Pitu cheese elaboration, the coagulation of cow's milk is achieved through acidification by lactic acid producing cultures and with little coagulating enzyme. The stability of κ-casein on the casein micelles surface is lowered either by "cutting" the hairs (enzyme action) or neutralizing the net charge of caseins (lowering pH), both effects playing synergistically a role. Thus, mainly hydrophobic and hydrogen bonding are important for the cheese network integrity [1]. These so-called "combined gels" can be considered as composite materials since they are composed of a continuous phase (protein matrix) formed by protein strands, and two dispersed phases (fat and moisture) as fillers in the matrix [2]. The rheological properties of cheese depend on the volume fraction of the dispersed phases, the shape, size distribution, and orientation of the dispersed units. Cryo-scanning electron microscopy (cryo-SEM) has facilitated the microscopical examination of dairy food high in water and/or fat with minimal distortion [3]. The objective of this work was to analyse the biochemical, microstructural and rheological properties of two varieties of Afuega'l Pitu cheese.

F. J. Galindo-Rosales et al. (Eds.): IBEREO 2019, SPM, pp. 54–58, 2020.
https://doi.org/10.1007/978-3-030-27701-7_12

2 Materials and Methods

2.1 Cheese Sampling

Afuega'l Pitu *atroncau* (truncated cone shape) *Blancu* (B) and *Roxu* (R) were manufactured in nine farms (B1−B9 and R1−R9) in Asturias (northern Spain) following the specifications of the Protected Designation of Origin (PDO). Nine cheeses of each variety were sent at 30 days of ripening and cut into halves: one for biochemical and SEM analysis, and the other one for the rheological tests.

2.2 Biochemical Analyses

Total solids (TS), protein, fat and pH determinations following the standards referenced by [4]. NaCl content was determined by the official AOAC method [5].

2.3 Cryo-SEM Analysis

Cheese blocks (2 × 1 × 5 mm) were cryo-fixed by plunging into nitrogen slush (−210 °C). Frozen specimens were transferred, under vacuum at −150 °C, to the preparation cryo-chamber. Specimens were fractured, etched (5 min, 90 °C), and then coated with Au/Pd by sputtering for 45 s with a 12 mA current. The sample was transferred into the SEM chamber and studied at −150 °C.

2.4 Stress Sweep Test

A stress-controlled rotational rheometer Physica MCR-301 (Anton Paar, Graz, Austria) with a PP10/P2 geometry (1 mm gap) was used. Samples were tempered for 15 min at the correspondent test temperature. Stress sweeps from 10 Pa to 10000 Pa at 1 Hz and 20 °C were applied to delimit the linear viscoelastic range by using complex modulus (G^*) with the range of tolerable deviation (±10%). Stress (σ_{max}) and strain (γ_{max}) amplitudes and loss factor ($tan\delta$) were calculated.

2.5 Statistical Analysis

ANOVA and Pearson correlation coefficients ($^{***}P < 0.001$, $^{**}P < 0.01$, $^{*}P < 0.05$) were carried out with Statistica 8.0 (Statsoft Inc., Tulsa, USA). Data were expressed as mean values and standard deviations of at least three replicate tests.

3 Results and Discussion

3.1 Cheese Composition

Table 1 shows the B1−B9 and R1−R9 compositional parameters. Significant differences ($P < 0.05$) were observed, specifically for pH, TS and NaCl content which may stem from different elaboration or ripening conditions (rennet dose, coagulation

Table 1. Afuega'l Pitu *atroncau Blancu* (B1–B9) and *Roxu* (R1–R9) biochemical parameters

	pH	Total solid, TS (g/100 g)	Fat (g/100 g TS)	Protein (g/100 g TS)	Fat-to-protein ratio, FPR	Moisture-to-protein ratio, MPR	NaCl (g/100 g TS)
B1	4.53 ± 0.02^{abA}	71.4 ± 0.3^{aA}	50 ± 1^{abA}	39.1 ± 0.5^{aA}	1.29 ± 0.05^{aceA}	1.03 ± 0.02^{aA}	1.5 ± 0.2^{abA}
B2	4.46 ± 0.01^{aC}	74.2 ± 0.6^{bC}	51 ± 2^{abB}	39.1 ± 0.7^{aB}	1.29 ± 0.06^{acB}	0.89 ± 0.04^{bC}	1.90 ± 0.01^{abcB}
B3	4.54 ± 0.03^{bE}	54.16 ± 0.00^{cE}	48.3 ± 0.6^{acC}	40.6 ± 0.2^{bC}	1.19 ± 0.02^{adC}	2.08 ± 0.01^{cE}	1.39 ± 0.03^{aC}
B4	4.28 ± 0.04^{cG}	68.4 ± 0.3^{dG}	51 ± 2^{abE}	36.2 ± 0.2^{cdE}	1.42 ± 0.05^{beE}	1.28 ± 0.01^{dG}	2.3 ± 0.3^{abcD}
B5	4.23 ± 0.02^{cH}	76.4 ± 0.7^{eI}	52 ± 1^{abF}	37.0 ± 0.2^{cF}	1.39 ± 0.04^{beF}	0.83 ± 0.03^{bH}	2.2 ± 0.3^{abcE}
B6	4.10 ± 0.00^{dI}	64.63 ± 0.03^{fJ}	50 ± 1^{abG}	37.0 ± 0.5^{cG}	1.36 ± 0.05^{ceG}	1.48 ± 0.02^{eI}	1.9 ± 0.4^{abcG}
B7	4.39 ± 0.01^{eK}	73.00 ± 0.08^{bL}	52.4 ± 0.5^{bH}	35.2 ± 0.2^{dI}	1.49 ± 0.02^{bH}	1.05 ± 0.01^{aK}	2.4 ± 0.4^{bcH}
B8	4.49 ± 0.01^{abM}	56.7 ± 0.2^{gN}	49 ± 1^{abcJ}	36.3 ± 0.5^{cdK}	1.36 ± 0.05^{ceJ}	2.10 ± 0.03^{cM}	2.67 ± 0.01^{cI}
B9	4.03 ± 0.01^{dO}	69.4 ± 0.1^{dP}	45.3 ± 0.5^{cK}	41.8 ± 0.8^{bL}	1.09 ± 0.03^{dK}	1.05 ± 0.02^{aN}	1.7 ± 0.1^{abK}
R1	4.29 ± 0.03^{abB}	69.31 ± 0.09^{aB}	51.7 ± 0.2^{aA}	39.2 ± 0.2^{aA}	1.32 ± 0.01^{aA}	1.13 ± 0.01^{aB}	1.73 ± 0.01^{aA}
R2	4.40 ± 0.00^{bD}	71.53 ± 0.08^{bD}	51 ± 1^{aB}	38.3 ± 0.5^{aeB}	1.33 ± 0.05^{aB}	1.04 ± 0.01^{bD}	1.8 ± 0.3^{aB}
R3	4.37 ± 0.04^{abF}	66.81 ± 0.04^{cF}	50.8 ± 0.2^{aD}	37.82 ± 0.05^{beD}	1.34 ± 0.01^{aD}	1.314 ± 0.002^{cF}	1.5 ± 0.2^{aC}
R4	4.18 ± 0.04^{cdG}	69.4 ± 0.4^{aH}	53 ± 2^{aE}	35.9 ± 0.2^{cE}	1.47 ± 0.05^{bE}	1.22 ± 0.02^{dG}	1.3 ± 0.1^{aD}
R5	4.14 ± 0.06^{dgH}	75.8 ± 0.2^{dI}	51.3 ± 0.4^{aF}	36.7 ± 0.2^{bcF}	1.40 ± 0.02^{abF}	0.87 ± 0.01^{eH}	1.22 ± 0.05^{aF}
R6	4.27 ± 0.01^{acJ}	65.17 ± 0.02^{eK}	49.6 ± 0.4^{aG}	38.5 ± 0.3^{aH}	1.29 ± 0.02^{aG}	1.39 ± 0.01^{cJ}	2.0 ± 0.3^{abG}
R7	4.49 ± 0.03^{eL}	74.33 ± 0.09^{fM}	49.8 ± 0.2^{aI}	36.2 ± 0.5^{cJ}	1.38 ± 0.03^{abI}	0.95 ± 0.02^{beL}	2.8 ± 0.3^{bH}
R8	4.62 ± 0.00^{fN}	54.2 ± 0.2^{gO}	49 ± 2^{abJ}	37.4 ± 0.7^{beK}	1.32 ± 0.07^{aJ}	2.26 ± 0.04^{fM}	1.24 ± 0.08^{aJ}
R9	4.06 ± 0.00^{gP}	70.5 ± 0.2^{hQ}	45.8 ± 0.7^{bK}	41.5 ± 0.1^{dL}	1.10 ± 0.02^{cK}	1.01 ± 0.01^{bN}	1.8 ± 0.1^{aK}

Values in the same column followed by different letters are significantly differences ($P < 0.05$): small letters for the same variety (*B* or *R*), and capital letters for different variety at the same farm (from 1 to 9)

temperature and time, kneading time, ripening conditions, etc.) [4]. B and R showed low salt content according to the values reported by [6] for different cheeses.

B9 and R9 showed low-fat content and the highest protein content, therefore the lowest fat-to-protein ratio (FPR). Conversely, B4, R4, B5, R5, and B7, R7 showed the greatest FPR. B5 and R5 exhibited the highest TS content and the lowest moisture-to-protein ratio (MPR), whereas B3 and B8, R8 presented the lowest TS content and the highest MPR.

3.2 SEM Images

B and R exhibited a packed protein matrix (continuous phase), comprising thick cross-linked strands formed by fused casein particles that occlude fat and water (dispersed phases) [1]. Nevertheless, two different network distributions were observed (Fig. 1): fat globules and protein matrix homogeneously distributed (B) *vs* protein matrix and larger fused-fat areas displayed (R). The latter might be due to the mechanical kneading process, proper to *Roxu* variety. This effect, along with the natural protein matrix constriction until 30d, exerted pressure on the cheese structure forcing fat to become

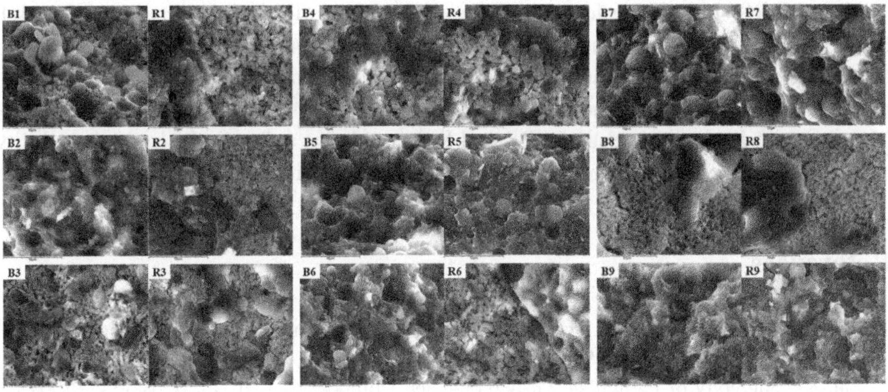

Fig. 1. Cryo-SEM images (5000x) of Afuega'l Pitu *Blancu* (B1–B9) and *Roxu* (R1–R9)

closer forming aggregates or non-globular fat. In addition, B2, R2, B5, R5, B7, R7, and B9, R9 showed a less porous protein matrix with thicker strands in agreement with their highest TS content (Table 1).

3.3 Linear Viscoelastic Range

B and R exhibited stiff structures with small γ_{max} (Table 2), as a result of the coarse structure typical of acid casein gels [7]. However, B exhibited higher γ_{max} and σ_{max} than R indicating more deformable and conformationally stable structures. Small fat globules (B) could produce a more flexible protein matrix with greater stability. Larger fat particles (R) disrupt the protein matrix [8] producing a more discontinuous and heterogeneous network with less structural stability. In general, *tanδ* similar values indicated a similar solid-like character in B and R cheese network.

Table 2. Viscoelastic parameters of Afuega'l Pitu *Blancu* (B1–B9) and *Roxu* (R1–R9)

	γ_{max} (%)	σ_{max} (Pa)	G^* (kPa)	$tan\delta$		γ_{max} (%)	σ_{max} (Pa)	G^* (kPa)	$tan\delta$
B1	0.16 ± 0.02^{aA}	886 ± 89^{aA}	570 ± 63^{aA}	0.277 ± 0.003^{aA}	R1	0.150 ± 0.004^{aA}	699 ± 70^{aB}	466 ± 13^{aB}	0.289 ± 0.002^{abB}
B2	0.29 ± 0.04^{bdB}	1255 ± 126^{bC}	441 ± 60^{bfC}	0.294 ± 0.004^{bC}	R2	0.19 ± 0.01^{abC}	997 ± 100^{bD}	536 ± 28^{aC}	0.289 ± 0.001^{abC}
B3	0.37 ± 0.01^{bcD}	630 ± 63^{cE}	172 ± 5^{ceD}	0.280 ± 0.001^{adD}	R3	0.22 ± 0.02^{bcE}	997 ± 100^{bF}	461 ± 38^{aE}	0.295 ± 0.001^{adE}
B4	0.32 ± 0.03^{bcdF}	998 ± 100^{dG}	316 ± 28^{bgF}	0.290 ± 0.003^{bcF}	R4	0.24 ± 0.04^{cF}	794 ± 79^{cH}	341 ± 56^{bF}	0.291 ± 0.006^{abF}
B5	0.35 ± 0.05^{bcG}	2506 ± 251^{eI}	735 ± 96^{dG}	0.287 ± 0.004^{cdG}	R5	0.28 ± 0.01^{cdH}	1991 ± 199^{dJ}	702 ± 24^{cG}	0.286 ± 0.003^{bG}
B6	0.39 ± 0.03^{cI}	999 ± 100^{dK}	260 ± 22^{cgH}	0.280 ± 0.001^{adH}	R6	0.22 ± 0.02^{bcJ}	630 ± 63^{eL}	293 ± 22^{bH}	0.286 ± 0.003^{bI}
B7	0.27 ± 0.03^{dK}	1258 ± 126^{bM}	474 ± 62^{afI}	0.281 ± 0.005^{adJ}	R7	0.150 ± 0.006^{aL}	1256 ± 126^{fM}	839 ± 33^{dJ}	0.257 ± 0.005^{cK}
B8	0.28 ± 0.01^{bdM}	251 ± 25^{fO}	88 ± 4^{eK}	0.293 ± 0.001^{bcL}	R8	0.29 ± 0.01^{dM}	251 ± 25^{gO}	87 ± 4^{eK}	0.301 ± 0.001^{dM}
B9	0.29 ± 0.02^{bdN}	1296 ± 130^{bP}	442 ± 4^{abfL}	0.296 ± 0.001^{bN}	R9	0.26 ± 0.02^{cdN}	1991 ± 199^{dQ}	777 ± 56^{cdM}	0.292 ± 0.002^{abdO}

$^{a\text{-}g}$ in the same column indicate significant differences ($P < 0.05$) for the same variety (*B* or *R*); $^{A\text{-}Q}$ in the same row for the same parameter indicate significant differences at the same trial (from 1 to 9)

B8 and R8 showed the lowest σ_{max} and G^* values (Table 2) in line with the high MPR (Table 1), provoking a connectivity reduction among protein strands and therefore in the number of intra- and inter-strands linkages. This trend was assessed by high negative correlations between MPR–σ_{max} (-0.754^{***}) and MPR–G^* (-0.859^{***}). Moreover, an intensive kneading (B8 and R8) could also enhance the hydration ability and consequently weakening the cheese network structure.

References

1. van Vliet, T., Lakemond, C., Visschers, R.: Rheology and structure of milk protein gels. Curr. Opin. Colloid Interface Sci. **9**, 298–304 (2004)
2. Pal, R.: Rheology of Particulate Dispersions and Composites. CRC Press, Florida (2007)
3. El-Bakry, M., Sheehan, J.: Analysing cheese microstructure: a review of recent developments. Food Eng. **125**, 84–96 (2014)
4. Piñeiro, L., Franco, I., Tovar, C.A.: The Multidisciplinary Science of Rheology. Towards a Healthy and Sustainable Development, pp. 56–59 (2017)
5. Horwitz, W.: Official Methods of Analysis of the Association of Official Analytical Chemists, 18th edn. AOAC Intel, Gaithersburg (2005)
6. Ramírez-Navas, J., Aguirre-Londoño, J., Aristizabal-Ferreira, V., Castro-Narváez, S.: La sal en el queso: diversas interacciones. Agron. Mesoam. **28**, 303–316 (2016)
7. Renkema, J.: Relations between rheological properties and network structure of soy protein gels. Food Hydrocoll. **18**, 39–47 (2004)
8. Piñeiro, L., Franco, I., Tovar, C.A.: Rheological and biochemical study of Afuega'l Pitu cheese (PDO). In: Hernández, M.J., Sanz, T., Salvador, A., Rubio-Hernández, F.J., Steinbrüggen, R. (eds.) The Multidisciplinary Science of Rheology. Towards a Healthy and Sustainable Development, pp. 56–59 (2017)

Rheological Approach in the Design of Adequate Multicomponent Food Ingredient System to Develop a Non-alcoholic Supercooled Beverage

Lídia Pinheiro[1(✉)], Catarina Maia[2], Catarina M. M. Duarte[3], Maria do Rosário Bronze[4], and Cátia Saldanha do Carmo[5]

[1] Research Institute for Medicines (iMED.ULisboa), Faculty of Pharmacy, Universidade de Lisboa, Lisbon, Portugal
lpinheiro@ff.ulisboa.pt
[2] iBET, Instituto de Biologia Experimental e Tecnológica, Oeiras, Portugal
[3] ITQB, Instituto de Tecnologia Química e Biológica António Xavier, Universidade Nova de Lisboa, Lisbon, Portugal
[4] iMED.Ulisboa, iBET and ITQB, Lisbon, Portugal
mbronze@ibet.pt
[5] iBET and ITQB, Lisbon, Portugal
catia.saldanhadocarmo@gmail.com

Abstract. The desired textural and rheological features of foodstuffs can be achieved through the addition of hydrocolloids, which increases viscosity and stability of food products. The investigation of rheological properties and dispersion stability of combinations of commonly used gums in the food industry for the development of a stable supercooled beverage was performed. Different formulations were tested, combining arabic gum (AG), xanthan gum (XG) and locust bean gum (LBG) at different gums concentrations and proportion of each ingredient. Results showed that all formulations were adequately described by power-law model. Moreover, the flow characteristics of the gums mixtures varied between shear-thinning and shear-thickening behavior. The apparent viscosity and the flow index were greatly affected by the total concentration of gums. Furthermore, XG was identified as an important ingredient for the formulation of non-alcoholic supercooled beverages concerning their stability. Response Surface Methodology (RSM) and Central Composite Rotatable Design (CCRD) were effective in optimizing the formulation parameters (apparent viscosity and zeta potential). A desirable higher viscosity and stable dispersion (lower zeta potential) under the temperature range 20 °C and −10 °C was achieved at a combined level of 0.08% w/w, 42.5% AG, 43.03% of XG and 14.47% of LBG.

1 Introduction

Hydrocolloids have been largely used in food industry as modifiers of appearance, texture and rheological properties because they can stabilize emulsions and dispersions [1]. Combining more than one type of hydrocolloids has been able to respond to

F. J. Galindo-Rosales et al. (Eds.): IBEREO 2019, SPM, pp. 59–63, 2020.
https://doi.org/10.1007/978-3-030-27701-7_13

increasingly demanding requests in food domains [1]. Thermal and physical treatments are usually applied to food products along the processing and distribution steps, affecting textural and rheological properties of hydrocolloid solutions. For this reason, it is detrimental to investigate the thermal and freezing properties of hydrocolloids to be used in food formulations [2]. For instance, the high affinity of hydrocolloids for water and their ability to increase the viscosity of mixtures determine the minimum size of ice crystals that can be achieved [3]. The major goal of this study was to evaluate the effect of hydrocolloids AG, XG and LBG on viscosity and stability of a non-carbonated beverage model (NCBM) to obtain a stable supercooled beverage (SSB), investigating also their synergistic effects. So far, there are no published works studying the rheological behavior of hydrocolloids in a non-alcoholic beverage model at different cooling temperatures. The similar studies found in literature are mostly dealing with stability of beverage emulsions containing hydrocolloids [4] or intrinsic viscosity and flow behaviour of aqueous solutions containing different hydrocolloids at much higher temperature ranges (invariably above 20 °C) [5]. A correlation between their concentrations and the flow properties of the NCBM under a temperature range of 20 °C and −10 °C was also set up. The present work evaluates, for the first time, the behavior of combinations of the mentioned commercial used gums in the food industry in a beverage model under different cooling temperatures to develop a SSB. The rheological properties at various temperatures and the dispersion stability of the model containing different proportions of gums were exhaustively determined.

2 Materials and Methods

2.1 Preparation of the NCBM Containing Polysaccharide Solutions

NCBM was formulated according to Dyrby et al., 2001, having a °Bx of 8 and a pH of approximately 3 [6]. Response Surface Methodology (RSM) was carried out following a Central Composite Rotatable Design (CCRD), as a function of three factors: Total concentration of gums (Σ [Gums] (% w/w), Proportion of AG (%) and Proportion of XG (%). The Σ [Gums] ranged from 0.038 to 0.122, the Proportion of AG varied from 30 to 65 and the Proportion of XG varied from 8.7 to 34.3, in accordance with the experimental design adopted. LBG was added to the non-carbonated beverage model by default.

2.2 Zeta Potential (ζ-Potential) Measurements

Electrophoretic-mobility (EM) measurements were conducted using the Malvern Zetasizer Nano ZS. Measurements of EM were performed to estimate the surface net charge of particles in suspension [7]. The ζ-potentials of samples were determined from the EM by enforcing the Henry equation. All measurements were done in triplicate and conducted at 25 °C, and at dispersions with pH 3 value.

2.3 Viscosity and Fluid Behavior Measurements

Characterization of viscosity behavior of systems composed by the NCBM with added gums during temperature changes was carried out using a programmable Brookfield LVDV II+Pro digital viscometer. Seven rotational speeds from 100 to 200 rpm and a sample volume of 45 mL were used. Spindle SC-18 was selected, together with a small sample adapter accessory connected to a refrigerated water bath. Solutions of gum samples were equilibrated at desired temperature before measurements.

2.4 Statistical Analysis

CCRD results were analyzed using the software Statistica™, version 10. Statistical analysis was carried out by analysis of variance (ANOVA) with $P < 0.05$.

3 Results and Discussion

Zeta potential varied from −7.1 and −30.2 mV amongst samples of the design, being equal to −2.8 mV for NCBM. The NCBM containing hydrocolloids was found to be negatively charged, which may be ascribed to the fact that XG is a negatively charged ionic polysaccharide [8]. The optimum region corresponded to the largest magnitude of negatively charged zeta potential (−30.2 and −30.0) resulting from a combined level of 0.122% w/w Σ [Gums], 65% proportion of AG, 8.7% proportion of XG and by default 26.3% LBG and a combined level of 0.08% w/w Σ [Gums], 42.5% proportion of AG, 43.03% proportion of XG and by default 14.47% LBG, respectively. Flow curves and viscosity profiles were investigated at different temperatures for all samples of the design. Shear stress (SS)-shear rate (SR) data were examined for Newtonian, Bingham, power-law, Herschel Bulkley and Casson rheological models. It was found that the curves fitted the power-law model appropriately. The analysis of viscosity profiles was important in order to determine their flow behavior and the influence of temperature on their apparent viscosity. In general, most formulations presented shear-thickening characteristics across the temperatures tested, with n (flow behavior index) >1. With decreasing temperatures, a sample containing the highest content of gums (0.151%) and similar proportions of AG, XG and LBG, exhibited a strong shear-thinning behavior ($n < 1$). Flow curves and viscosity profile of a sample produced at optimum conditions are presented in Fig. 1. A very large shear-thinning behavior was observed at −8 and −10 °C, probably due to high content in XG that is very sensitive to shear, implying a predisposition of XG long chain molecules to orientate along the flow direction.

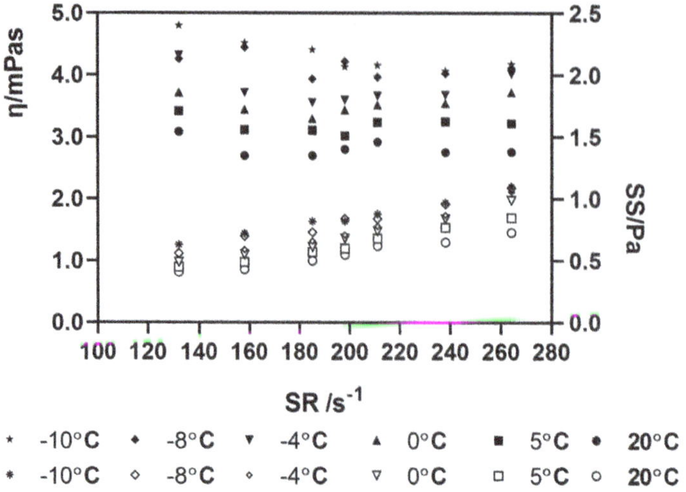

Fig. 1. Viscosity profile and flow curves for optimal sample (Σ [Gums] = 0.08% w/w, AG = 42.5%, XG = 43.0% and LBG = 14.5%); SS – Shear stress; SR – Shear rate

4 Conclusions

Total gums concentration contributed significantly to the rheology of the selected gums in the non-carbonated supercooled beverage. The flow behavior was well described by the power-law model and gums mixtures behaved like shear-thinning or shear-thickening fluids depending firstly on concentration of gums and secondly on proportion of XG. The results clearly highlight XG as an important ingredient for the formulation of non-alcoholic supercooled beverages concerning their stability. A desirable higher viscosity and stable dispersion under the temperature range 20 °C and −10 °C was achieved from the overall optimum domain at a combined level of 0.08% w/w, 42.5% of AG, 43.03% of XG and 14.47% of LBG.

References

1. Pinheiro, A.C., Bourbon, A.I., Rocha, C., Ribeiro, C., Maia, J.M., Gonçalves, M.P., Teixeira, J.A., Vicente, A.A.: Rheological characterization of κ-carrageenan/galactomannan and xanthan/galactomannan gels: Comparison of galactomannans from non-traditional sources with conventional galactomannans. Carbohydr. Polym. **83**(2), 392–399 (2011)
2. Zameni, A., Kashaninejad, M., Aalami, M., Salehi, F.: Effect of thermal and freezing treatments on rheological, textural and color properties of basil seed gum. J. Food Sci. Technol. **52**(9), 5914–5921 (2014)
3. Bezerra, T.S., Fernandes, T.N., de Resende, J.V.: Effects of added sucrose and pectin on the rheological behavior and freezing kinetics of passion fruit pulp Studied by response surface methodology. J. Food Sci. Technol. **52**(6), 3350–3357 (2015)

4. Mirhosseini, H., Tan, C.P., Hamid, N.S.A., Yusof, S.: Modeling the relationship between the main emulsion components and stability, viscosity, fluid behavior, ζ-potential, and electrophoretic mobility of orange beverage emulsion using response surface methodology. J. Agric. Food Chem. **55**, 7659–7666 (2007)
5. Dyrby, M., Westergaard, N., Stapelfeldt, H.: Light and heat sensitivity of red cabbage extract in soft drink model systems. Food Chem. **72**(4), 431–437 (2001)
6. Mirhosseini, H., Tan, C.P., Taherian, A.R., Boo, H.C.: Modeling the physicochemical properties of orange beverage emulsion as function of main emulsion components using response surface methodology. Carbohydr. Polym. **75**(3), 512–520 (2009)
7. Gómez-Díaz, D., Navaza, J.M., Quintáns-Riveiro, L.C.: Intrinsic viscosity and flow behaviour of Arabic gum aqueous solutions. Int. J. Food Prop. **11**(4), 773–780 (2008)
8. Sun, C., Gunasekaran, S., Richards, M.P.: Effect of xanthan gum on physicochemical properties of whey protein isolate stabilized oil-in-water emulsions. Food Hydrocoll. **21**(4), 555–564 (2007)

Rheological Evaluation of Ethyl Cellulose and Beeswax Oleogels as Fat Replacers in Meat Products

Beatriz Herranz[1(✉)], Susana Cofrades[2], Joaquín Gómez-Estaca[2], and María Dolores Álvarez[1]

[1] Department of Characterization, Quality and Safety, Institute of Food Science, Technology and Nutrition (ICTAN-CSIC), José Antonio Novais 10, 28040 Madrid, Spain
{beatriz.herranz,mayoyes}@ictan.csic.es
[2] Department of Products, Institute of Food Science, Technology and Nutrition (ICTAN-CSIC), Madrid, Spain
{scofrades,jgomez}@ictan.csic.es

Abstract. Ethyl cellulose and beeswax oleogels were characterize physico-chemical and rheological to study their suitability to be employed as fat replacers for healthier meat products. The effect of storage time during 28 days was also evaluated. Physicochemical (color, fatty acid composition, lipid oxidation), mechanical (puncture) and viscoelastic properties of both oleogeles were measured at 1 and 28 days. Ethyl cellulose produced more deformable and cohesive gels with greater time- and temperature- stability than their beeswax counterparts which resulted in rigid and brittle gels. Their oxidative stability, mechanical and rheological properties were very stable during the chilled storage period, suggesting that both organogels could be stored up to 15 days prior to use without significant change in their composition or technological properties.

1 Introduction

Rising consumer concern about diet and health has prompted the development of healthier meat products based on their lipid composition [1]. In order to reduce fat or cholesterol or improve fatty acid profile, reformulation generally entails replacement of the animal fat present in traditional products with another lipid more in line with health recommendations such as vegetable or marine oils [1, 2]. However, the meat products formulated with alternative lipid materials have different physico-chemical characteristics than their traditional counterparts, resulting in a negative effect on the desired quality attributes of the reformulated products [2]. A possible way to overcome the quality issues associated with animal fat substitution in meat products is the structuring of liquid lipids into soft matter structures with solid-lipid functionality similar to that of animal fat. Among organogelation systems for meat products development, ethyl cellulose gels and beeswax stand out because of the possibility of textural modifications to resemble the properties of traditional products. The objective of the present work

F. J. Galindo-Rosales et al. (Eds.): IBEREO 2019, SPM, pp. 64–68, 2020.
https://doi.org/10.1007/978-3-030-27701-7_14

was to develop and characterize rheological animal fat replacers by structuring a healthier lipid mixture (olive, linseed, fish) with two different organogelators (ethyl cellulose and beeswax), taking into consideration their physico-chemical, mechanical (puncture) and viscoelastic properties as a function of storage (28 days, 3 ± 1 °C).

2 Materials and Methods

2.1 Oleogel Development

Ethyl cellulose (EC-OG) and beeswax oleogels (W-OG) were made as described by Gómez-Estaca and col. [3]. Their attributes were evaluated at 1 day and after 28 days of refrigerated storage (3 ± 1 °C).

2.2 Fatty Acid Composition and Lipid Oxidation (TBARS)

For fatty acid composition, ten mg of the oil mixture or oleogel were derivatized into fatty acid methyl esters, which were determined in an Agilent 7820A gas chromatograph with FID detector as described by Gómez-Estaca and col. [3]. Results were expressed as mg fatty acid/g oil.

TBARS were determined as described in previous work [3]. Results were expressed as mg malonaldehyde (MDA)/kg oil.

2.3 Texture Analysis

Six replications of a penetration test were performed in a TA-XTplus Texture Analyzer. The parameters obtained from the corresponding force–distance curves were penetration force (N) at 10 mm and gel strength (N mm), defined as the area below the force-distance curve.

2.4 Small Amplitude Oscillatory Strain (SAOS) Measurements

They were performed using a Kinexus pro rheometer. Amplitude sweeps were run at 1 Hz by varying the shear strain (γ) from 0.01 and 0.001 up to 10% for EC-OG and W-OG, respectively. Frequency sweeps were run at from 0.01 to 10 Hz. The strain amplitude was set at 0.1 and 0.01%, within the LVE range for EC-OG and W-OG respectively.

3 Results and Discussion

3.1 Fatty Acid Composition and Lipid Oxidation (TBARS)

The fatty acid composition of the oil mixture was quite similar (data not shown) to that previously reported by Delgado-Pando et al. [4] showing 15.5% ΣSFA, 48.5% ΣMUFA and 36% ΣPUFA according to the composition of oils in the mixture (olive, lineseed and fish). The ΣPUFA/ΣSFA ratio decreased significantly for both oleogels, especially for EC-OG. In contrast, the ΣPUFAn-6/ΣPUFAn-3 ratio remained unchanged.

Both gel systems showed a significantly ($p \leq 0.05$) higher TBARS value at 1 day (data not shown), suggesting the oxidation of lipids as a consequence of the oleogel manufacturing process. This is especially true for EC-OG attaining higher TBARS values because of the high processing temperatures needed to unfold EC for gel formotion, in contrast to W-OG which is produced at 65 °C. W-OG remained relative-y stable up to 2 weeks after production. EC-OG showed a similar trend observing an increase in TBARS values was significant at the 4th week of storage.

3.2 Mechanical and Rheological Properties

With respect to the mechanical parameters evaluated, W-OG attained a significantly ($p \leq 0.05$) higher penetration force and gel strength values than EC-OG (Table 1). After 28 days of chilled storage, neither the mechanical profile nor maximum force or gel strength had changed), irrespective of the type of organogelator used.

Table 1. Penetration test parameters of EC-OG and W-OG at 1 day and 28 days of chilled storage

	Day	Penetration force (N)	Gel strength (N mm)
EC-OG	1	6.5 ± 0.8a	57.3 ± 4.7a
	28	6.7 ± 0.70a	59.0 ± 6.9a
W-OG	1	11.7 ± 2.3b	99.9 ± 17.5b
	28	11.9 ± 0.8b	95.7 ± 5.2b

Different letters in the same column indicate significant differences ($p \leq 0.05$).

As can be seen in Table 2, beeswax gels showed significantly higher and lower values ($p < 0.05$) of complex modulus (G*) and γ_{max}, respectively, than those of EC (Table 1). This indicates that beeswax produced denser but less deformable gels than those produced with EC. Hence, the beeswax molecule seems to reinforce the oleogel matrix, increasing the crosslink density in the oil system and making more rigid, stronger and less deformable oleogels as compared with EC. EC is a polymer network held together by physical bonds, mainly hydrogen bonds, while in the case of beeswax the fatty acids and triglycerides assemble into a crystal by van der Waals interactions, making a particle network [5]. A less bulky conformation, as in the case of beeswax, allows the approach of molecules that can interact more closely with their neighbors through short-range interactions, leading to stronger, elastic structures [5]. Therefore the chemical nature of beeswax produces a structure with greater crosslink density, enabling a larger volume of the mobile oil phase to become trapped as compared to ethyl cellulose. No significant differences ($p < 0.05$) were found as function of storage time for stress and strain amplitude and G*, so both oleogel seem to be quite time-stable.

Table 2. Parameters from linear viscoelastic range of EC-OG and W-OG at 1 day and 28 days of chilled storage

	σ_{max} (Pa)	γ_{max} (%)	G* (kPa)	tan δ
W-OG-1 day	217 ± 15a	0.031 ± 0.0002b	694.0 ± 15.2a	0.199 ± 0.015b
W-OG-28 days	169 ± 12a	0.031 ± 0.0002b	534.3 ± 29.5a	0.168 ± 0.007c
EC-OG-1 day	149 ± 4a	0.398 ± 0.0003a	37.4 ± 1.3b	0.228 ± 0.0003a
EC-OG-28 days	148 ± 5a	0.399 ± 0.001a	37.2 ± 2.5b	0.222 ± 0.004a,b

Different letters in the same column indicate significant differences ($p \leq 0.05$).

Regarding the mechanical oleogel spectra (Fig. 1), both organogels behaved as solid-like materials (*G' > G"*) irrespective of frequency, While *G'* moduli was practically frequency-independent, *G"* increased slightly at low frequencies (0.01–0.1 Hz), mainly in the W-OG samples, reflecting the weak gel character. The differences between the two viscoelastic moduli were less than one order of magnitude in both organogels, although in the case of W-OG both moduli were greater than those of EC-OG so the network formed in the W-OG samples is rigid and at the same time more sensitive to shear [6]. This result could also indicate that, to some extent, this type of network could be easily damaged. The mechanical profiles of both oleogels at 28 days were practically the same as at day 1 (Fig. 1).

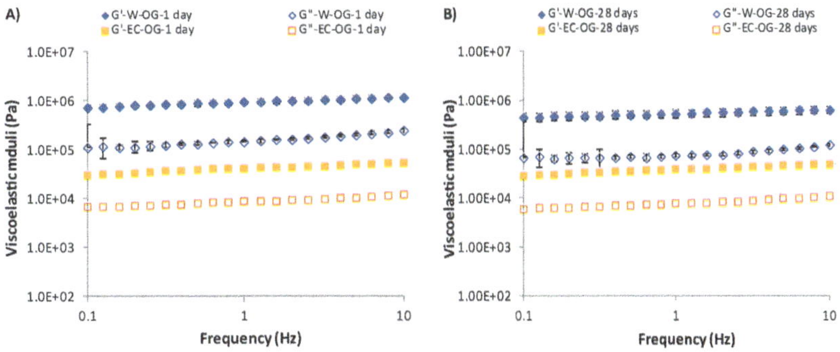

Fig. 1. Mechanical spectra of EC-OG and W-OG at 1 (A) and 28 days (B) of chilled storage. T = 25 °C. Closed symbols, storage modulus (*G'*); open symbols, viscous modulus (*G''*).

4 Conclusions

Both oleogels showed a solid-like structure similar to that of animal fat suggesting their potential use as fat replacers, with a healthy fatty acid profile in meat products development. EC-OG produced more deformable and cohesive gels with greater time stability than W-OG. Both oleogles could be stored for up to 15 days prior to use without significant changes in their composition or technological characteristics.

References

1. Jimenez-Colmenero, F., Salcedo-Sandoval, L., Bou, R., Cofrades, S., Herrero, A.M., Ruiz-Capillas, C.: Trends Food Sci. Technol. **44**, 177–188 (2015)
2. Barbut, S., Wood, J., Marangoni, A.: Meat Sci. **122**, 155–162 (2016)
3. Gómez-Estaca, J., Pintado, T., Jiménez-Colmenero, F., Cofrades, S.: Food Bioprocess Tech. **12**, 1068–1081 (2019)
4. Delgado-Pando, G., Cofrades, S., Ruiz Capillas, C., Solas, M., Jiménez-Colmenero, F.: Eur. J. Lipid Sci. Technol. **112**, 791–801 (2010)
5. Cerqueira, M.A., Fasolin, L.H., Picone, C.S.F., Pastrana, L.M., Cunha, R.L., Vicente, A.A.: Food Res. Int. **96**, 161–170 (2017)
6. Herranz, B., Tovar, C.A., Borderias, A.J., Moreno, H.M.: Innov. Food Sci. Emerg. Technol **20**, 24–33 (2013)

An Objective and Subjective Characterization of the Oral Processing of Six Solid Foods

María Dolores Álvarez[1(✉)], Jaime Paniagua[2], and Beatriz Herranz[1]

[1] Department of Characterization, Quality, and Safety, Institute of Food Science, Technology and Nutrition (ICTAN-CSIC), José Antonio Novais 10, 28040 Madrid, Spain
{mayoyes, beatriz.herranz}@ictan.csic.es
[2] Foundation Pita López, Matalpino s/n, 28400 Madrid, Spain
logocerebral@gmail.com

Abstract. Research of food oral processing is turning into progressively more needed with the increase of elderly people around the world. A new strategy for instrumental assessment of texture perception is needed, being also desirable to measure dynamic bolus formation and rheological properties for a better understanding. Banana, apple, carrot, cured ham, peanut and potato chips are solid foods with very different texture. Kramer test, performed with a miniature cell, could be very appropriate to establish an instrumental standardized approach of solid foods texture evaluation, and even of bolus counterparts, enables to know how foods mechanical properties change during dynamic oral processing.

1 Introduction

Food oral processing is an essential and complex process narrowly related to foods sensory perception, especially to texture perception [1]. In turn, texture perception is a dynamic mechanism which depends on food properties such as composition, structure, and changes which take place during oral processing [2]. Miniature Kramer cell close up mimic the early stage of mastication. To our knowledge, there is no literature relative to its use for objective measures of food texture, and still few researches relate bolus mechanical properties to dynamic texture perception.

The aim of this work was to carry out a comprehensive study of the texture of six solid foods (banana, apple, carrot, cured ham, peanut and potato chips). Firstly, the mechanical properties of solid foods were measured using a miniature Kramer cell. In carrot, peanut and potato chips, acoustic properties were also recorded. Secondly, the dynamic rheological properties of boluses were measured. Thirdly, a sensory evaluation was performed by an untrained panel, and involved measures of oral physiological parameters (chewing duration, chews number, chew rate and average eating rate). Pearson correlations were established to find relationships between foods and boluses mechanical properties and texture perception.

F. J. Galindo-Rosales et al. (Eds.): IBEREO 2019, SPM, pp. 69–73, 2020.
https://doi.org/10.1007/978-3-030-27701-7_15

2 Materials and Methods

2.1 Solid Food Items

Canarian banana, *Golden Delicious* apple, *Nantesa* packaged carrot (Horticola ES-MA, S.L., Valladolid, Spain), dices of Serrano cured ham (Incarlopsa, Cuenca, Spain), fried peanut (Importaco Casa Pons Sa, Valencia, Spain) and potato chips (CylIber-snacks S.L., Valladolid, Spain) were acquired from a local supermarket (Mercadona, Madrid, Spain).

2.2 Mechanical Properties of Solid Foods and Boluses

Mechanical properties of solid foods were measured using a TA.HDPlus Texture Analyser Stable Micro Systems Ltd., Godalming, UK) equipped with a 250 kg load cell. Objective measurements were carried out by using a miniature Kramer shear (HDP/MK05) cell at deformation rate of 2 mm/s. A constant food volume fixed at $\approx 5.20\ cm^3$ was used. In carrot, peanut and potato chips, simultaneously with the force, the sound emitted during Kramer test was also recorded with an acoustic envelope detector (AED) [3].

Rheological measurements of boluses were carried out using a rotational Kinexus pro rheometer (Malvern Instruments Ltd., Worcestershire, UK), which was equipped with a 40 mm parallel–plate geometry (1-mm gap) for measuring banana, apple and potato chips boluses and a 20 mm parallel–plate geometry (1.5-mm gap) for measuring carrot, cured ham and peanut boluses. Temperature was kept at 37 °C. To determine the linear viscoelastic (LVE) region, strain amplitude sweeps were run at 1 Hz by varying the shear strain (γ) from 0.01 up to 10%. Frequency sweeps were run subjecting boluses to stress that varied harmonically with time from 0.1 to 50 Hz and γ = 0.01%. All tests were carried out at least in quintuplicate.

2.3 Sensory Analysis

Sensory evaluation was performed by 39 participants completing several trials/tasks of the six foods corresponding to four stages of oral processing: non-oral evaluation, first bite, chewing process determining eating behaviors, and bolus characterization. During either non-evaluation or chewing process, participants selected from a list given the texture adjectives that they considered applicable for describing the expected and perceived texture of each food item. Chew rate and average eating rate were calculated in accordance with Wee et al. [4].

3 Results and Discussion

3.1 Mechanical/Acoustical Properties of Solid Foods

Great differences were observed in the shape of the Kramer force-distance curves depending on the food tested (data not shown). Carrot had significantly ($P < 0.05$) the highest Kramer forces, followed by peanut and cured ham, potato chips, apple and banana, in that order (Table 1). Carrot also required the highest work, and both carrot and potato chips had similar SPL_{max} and average drop off values.

3.2 Rheological Properties of Boluses

Rheology of spat-out boluses is shown in Table 2. Carrot bolus had the highest critical shear stress (σ_{max}) and complex modulus (G^*_{max}) and the lowest shear strain amplitude (γ_{max}), reflecting a denser with higher rigidity but less flexible physical network. On the contrary, cured ham bolus had the highest conformational flexibility (γ_{max}value) and the lowest structural complexity (tan δ value closer to 1). Boluses studied all exhibited weak gel properties, but the gel structure was weaker in banana and cured ham boluses.

Table 1. Kramer mechanical properties of solid foods and AED parameters

Food item	Maximum force (N)	Average force (N)	Work (J)	SPL_{max} (dB)	Average drop off (dB)
Banana	23.4d (1.53)	8.08c (0.591)	0.195e (0.017)	-	-
Apple	83.8d (4.87)	31.1c (0.450)	0.779d (0.011)	-	-
Carrot	725a (42.3)	377a (24.5)	5.65a (0.368)	90.0a (5.50)	7.94a (0.814)
Cured ham	554b (71.4)	201b (21.0)	5.03b (0.525)	-	-
Peanut	574b (16.0)	200b (23.8)	2.48c (0.310)	80.6b (3.97)	6.11b (0.823)
Potato chips	244c (0.091)	36.8c (3.63)	0.921d (0.091)	86.9a,b(1.75)	7.25a,b (0.450)

Means (standard deviation, SD).

Table 2. Viscoelastic properties of boluses (limit values of LVE range at 1 Hz)

Food item	σ_{max} (kPa)	γ_{max} (%)	G^*_{max} (kPa)	tan δ (-)	G' (kPa)	G'' (kPa)	η^* (kPa s)
Banana	0.006e (0.001)	0.251c (0.001)	2.44c (0.548)	0.285a,b (0.016)	2.63d (0.140)	0.657d (0.040)	0.432d (0.023)
Apple	0.029e (0.006)	0.159d (0.001)	18.2c (3.58)	0.177d (0.010)	44.3b,c (6.31)	7.19b,c (0.920)	7.14b,c (1.01)
Carrot	0.222a (0.008)	0.100e (0.001)	222[a] (9.45)	0.197c,d (0.005)	166a (13.5)	28.7a (3.94)	26.7a (2.21)
Cured ham	0.135c (0.005)	1.00a (0.001)	13.4c (0.550)	0.292a (0.023)	10.7d (1.19)	2.65c,d (0.254)	1.75d (0.193)
Peanut	0.175b (0.021)	0.158d (0.000)	111b (12.9)	0.216c (0.002)	61.9b (5.21)	10.9b (1.34)	10.0b (0.853)
Potato chips	0.080d (0.010)	0.395b (0.000)	20.2c (2.45)	0.255b (0.004)	32.6c (3.44)	6.33b,c (0.593)	5.28c (0.555)

Means (standard deviation, SD).

σ_{max}: maximum shear stress; γ_{max}: maximum shear strain; G^*_{max}: maximum complex modulus; tanδ: loss tangent; G': elastic modulus; G'': viscous modulus; η^*: complex viscosity.

3.3 Sensory Analysis

Potato chips were preferred to be eaten in first place. Main drivers of product choice were lifestyle, time of the day, appearance and expected oral sensations. However, the texture had little influence on foods choices before tasting. At first bite, carrot and banana were scored with the highest and the lowest perceived force and loudness degree, respectively. Saliva incorporation was estimated by subtraction between weight of bolus after mastication and that of food sample. In banana and apple, with higher water content, had no saliva incorporation (Table 3), whereas the contrary was true in carrot, cured ham, peanut and potato chips. Potato chips bolus had the highest saliva incorporation. Chewing time and chews number were longer and higher in carrot and cured ham. Banana and apple were the faster foods, while cured ham was the slowest one. Positive and significant correlations were found between the texture expectations before consumption, and textural attributes perceived by the participants during mastication. As an example, Fig. 1 shows the frequency of occurrence of the textural attributes selected before and after consumption for banana. By considering the six foods, correlations ranging between $r = 0.823$ (for expected and perceived hardness) and $r = 0.993$ (for expected and perceived gumminess) were found. On the other hand, carrot bolus was perceived as the least consistent and adhesive, whereas banana and cured ham boluses had the highest degrees of adhesiveness and consistency, respectively.

Table 3. Oral physiological parameters from the chewing processes of the six solid foods

Food item	Saliva incorporation (g)	Chewing duration (s)	Chews number (-)	Chew rate (chews per s)	Eatingrate (g min^{-1})
Banana	−0.121b,c (1.40)*	10.0b (3.76)	11.4b (4.70)	1.16a (0.319)	43.8a (14.0)
Apple	−0.479c (1.73)*	10.6b (4.06)	14.1b (5.25)	1.34a (0.263)	49.9a (18.2)
Carrot	0.518a,b (1.58)	19.8a (6.87)	25.3a (10.6)	1.29a (0.313)	14.2b (4.81)
Cured ham	1.22a (0.723)	19.3a (8.06)	24.6a (10.5)	1.29a (0.280)	8.50b (3.80)
Peanut	0.759a (0.882)	10.1b (3.43)	12.4b (3.94)	1.30a (0.398)	10.0b (3.82)
Potato chips	1.28a (1.00)	11.0b (3.97)	13.3b (4.33)	1.27a (0.348)	11.4b (4.82)

Means (standard deviation, SD). *There was no saliva incorporation after mastication.

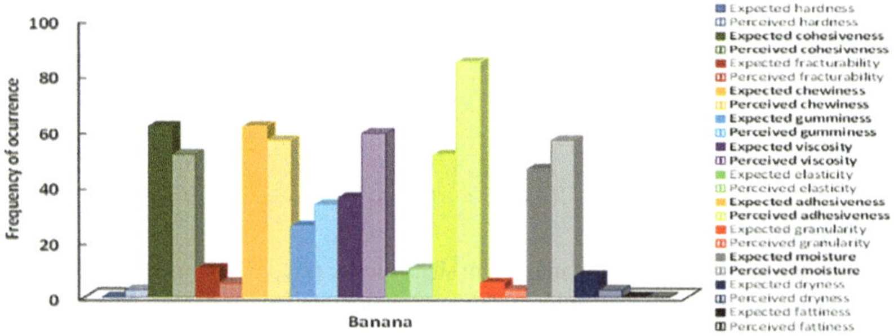

Fig. 1. Frequency of occurrence of expected and perceived textural attributes for banana

4 Conclusions

Kramer work would appear to be the best mechanical property for measuring the initial degree of structuring of the sixfoods studied. Critical stress (σ_{max}), representing bolus crosslink density, was highly correlated with Kramer mechanical properties, reflecting that a higher degree of structure in food is associated with a higher bolus network density. Higher foods mechanical properties are linked to longer chewing duration, greater chews number and slower eating rate. Perceived hardness had positive correlations with both Kramer forces, σ_{max}, G^*_{max}, G', G'' and η^*, and negative correlation with bolus adhesiveness, reflecting the interplay between texture perception, oral physiology and food properties.

References

1. Chen, J.: Food oral processing—a review. Food Hydrocoll. **23**, 1–25 (2009)
2. Hutchings, J.B., Lillford, P.J.: The perception of food texture—the philosophy of the breakdown path. J. Texture Stud. **19**, 103–115 (1988)
3. Salvador, A., Varela, P., Sanz, T., Fiszman, S.M.: Understanding potato chips crispy texture by simultaneous fracture and acoustic measurements, and sensory analysis. LWT - Food Sci. Technol. **42**, 763–767 (2009)
4. Wee, M.S.M., Goh, A.T., Stieger, M., Forde, C.G.: Correlation of instrumental texture properties from textural profile analysis (TPA) with eating behaviours and macronutrient composition for a wide range of solid foods. Food Funct. **9**, 5301–5312 (2018)

Rheology of Bioactive Hydrogels Formulated with Valuable Fractions from Discarded Potatoes

M. D. Torres[1(✉)], P. Fradinho[2], and H. Domínguez[1]

[1] Department of Chemical Engineering, Science Faculty, University of Vigo, Campus Ourense, Ourense, Spain
{matorres,herminia}@uvigo.es
[2] LEAF - Linking Landscape, Environment, Agriculture and Food, Instituto Superior de Agronomia, Universidade de Lisboa, Lisbon, Portugal
pfradinho@isa.ulisboa.pt

Abstract. Alternative functional hydrogels were formulated using gelling and bioactive extracts recovered from Galician discarded potatoes. Environmentally friendly technologies were used to extract the starch from the pulp and the bioactive fractions from the peels required to prepare the proposed hydrogels. The corresponding mechanical properties in terms of rheology and texture were carefully determined. The mechanical experiments indicated that these potato starchy hydrogels (20% w/w, 60–90 °C) exhibited adequate characteristics not only for food, but also for non-food applications. These innovative gelled systems could alleviate the current growing demand for functional hydrogels, adding value to residual local natural sources.

1 Introduction

In recent years, the recovery of biopolymers with gelling or bioactive properties from renewable sources have gained increased attention [1]. Potato could be an interesting alternative due to the high discarding degree (e.g. low sizes, damage or long storage samples, among others) and the simple extraction procedures when compared with other cereals [2]. Green extraction technologies, using only water as solvent, are currently preferred [3]. The formulation of functional hydrogels from residual fractions could be an attractive alternative to add value to these wastes, with the consequent advantages from the environmental, economic or social point of views.

The knowledge of the mechanical properties of the starchy-based hydrogels during processing is essential to select the most adequate final applications [4]. This behaviour is especially important in the case of functional hydrogels, where the processing conditions needs to be carefully controlled to achieve hydrogels with suitable rheological and textural properties, without jeopardising the bioactive characteristics of the incorporated functional compounds [5]. A number of studies have been published on the last years on the thermo-rheological features of potato gelled systems formulated with native [6] and modified potato starch [7], however no comprehensive works have

F. J. Galindo-Rosales et al. (Eds.): IBEREO 2019, SPM, pp. 74–78, 2020.
https://doi.org/10.1007/978-3-030-27701-7_16

been found on the mechanical properties of functional hydrogels prepared from extracts of disposal sources.

In this context, the main aim of this work was to study the rheological and textural features of functional potato starch-based hydrogels incorporated with bioactive compounds recovered from the peels of discarded potatoes.

2 Materials and Methods

2.1 Raw Materials

Three local waste potato varieties with low sizes (Kennebec, Agria, Neiker) were gently provided by INORDE (Instituto Ourensán de Desenvolvemento Económico, Galicia, Spain) and employed as raw material. The Galician varieties used to prepare the functional hydrogels were selected based on their starch and bioactives yield.

2.2 High Valuable Compounds Extraction

The starch isolation process from the pulp of three tested discarding potatoes was optimized using water as only extraction reagent, following the procedure previously detailed [8]. Extracted potato starch was stored in closed plastic boats at room temperature until further analysis. The extraction of bioactives from the corresponding peels of Neiker was carried out by autohydrolysis at 170 °C in a pressurized reactor (Parr Instruments series 4848, Illinois, USA), according to the protocol previously explained [8]. The liquid fraction was stored in the fridge until the hydrogels formulation (before one week).

2.3 Hydrogels Preparation

Functional hydrogels were formulated with the tested starches (60–90 °C) following the gelling conditions previously optimized for potato starch-based matrices in the absence of bioactives [2]. Bioactive liquid phases recovered from Neiker potato peels were used as solvent to prepare the hydrogels. Using both the Neiker liquor and the starch extracted (20%, w/w) from the three tested varieties, hydrogels were prepared at least in triplicate. Hydrogels were stored in the fridge 24 h previously to mechanical measurements in order to allow full gels maturation.

2.4 Mechanical Characterization

Viscoelastic behavior (G' and G") was determined in a stress-controlled rheometer (MCR302, Anton PaarPhysica, Austria). For this purpose, a plate-plate geometry (1 mm gap, 25 mm diameter) was used. Hydrogels previously formulated were placed on the plated measuring system, the exposed edges covered with paraffin oil to prevent water evaporation during measurements, and samples rested 10 min to allow thermal and structural equilibration of the samples. Viscoelastic properties of above hydrogels were conducted within the linear viscoelastic region (15 Pa, 5 °C) at least in triplicate, after performing the corresponding stress sweeps. Texture experiments were made at least in fivefold in a texture analyzer (TA-XT2, Stable MicroSystems, UK) using the texture profile analysis (TPA). A cylindrical probe P/05R (5 mm penetration and 1 mm/s crosshead speed) was employed. Note here that the syneresis of tested hydrogels two week aged was also analyzed following a standard procedure [9].

3 Results and Discussion

Figure 1 shows a representative example of functional hydrogels (20%, w/w) formulated with three tested discarded potato varieties at the largest gelling temperature (90 °C). Note here that lower gelling temperatures exhibited similar profiles with lower values of the elastic and viscous moduli (about 3-folds at 70 °C and 5-folds at 60 °C).

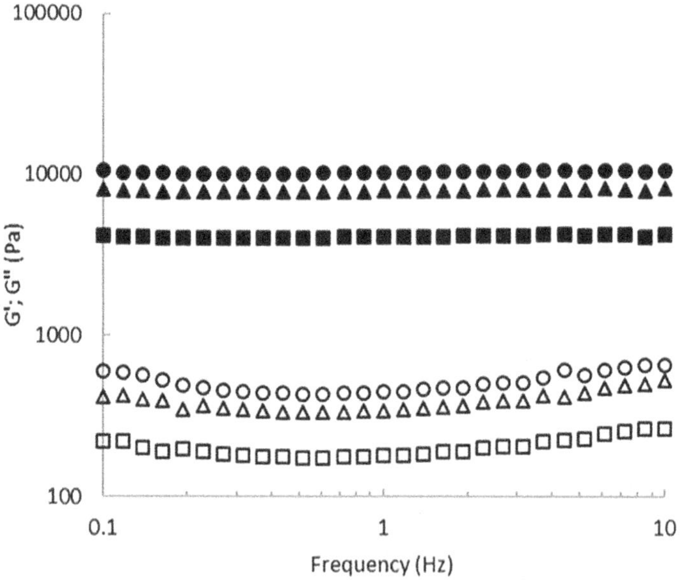

Fig. 1. Functional hydrogels (20% w/w) prepared with starches from Agria (circles), Kennebec (triangles) and Neiker (squares) and bioactive autohydrolysis liquor fractions from Neiker peels. Symbols: G', closed and G'', open.

In all cases, a typical gel behavior (G' > G'' and both moduli almost frequency independent) with intermediate gel strength can be clearly observed. It should be highlighted that those hydrogels prepared with Agria-starchy matrices presented the strongest gelling features followed by Kennebec and Neiker varieties. This behavior is consistent with the bioactive compounds content reported for each variety, increasing with decreasing gel strength [8]. Table 1 collects the textural parameters (firmness, adhesiveness, cohesiveness) of above functional hydrogels.

Table 1. Textural features of tested functional hydrogels

Starchy-based matrix	Firmness (N)	Adhesiveness (Ns)	Cohesiveness
Agria	0.95 ± 0.07	0.39 ± 0.05	0.29 ± 0.02
Kennebec	0.74 ± 0.05	0.28 ± 0.02	0.21 ± 0.02
Neiker	0.55 ± 0.03	0.19 ± 0.03	0.14 ± 0.03

Firmness, adhesiveness and cohesiveness exhibited the largest values for Agria variety. All hydrogels showed textural features with those observed in the rheological tests.

4 Conclusions

It should be concluded that rheological, textural and antioxidant experiments indicated that prepared functional starchy hydrogels featured attractive mechanical and bioactive properties, with promising food and non-food applications. The extraction of the starch from the discarded potatoes pulp and bioactive compounds from the corresponding peel could providing added value to these disposal fractions, being the optimization of the processing conditions critically relevant. A wide range of bioactive hydrogels could be achieved using the liquid extracts obtained by autohydrolysis with water as the only solvent for the hydrogels preparation, without jeopardising the mechanical properties.

Acknowledgements. Authors acknowledge the financial support (INOU 2018) to the Diputación Provincial de Ourense and University of Vigo. M.D. Torres thanks the Ministerio de Economía, Industria y Competitividad for her postdoctoral grant (IJCI-2016-27535).

References

1. Flórez-Fernández, N., Torres, M.D., González-Muñoz, M.J., Domínguez, H.: Potential of intensification techniques for the extraction and depolymerization of fucoidan. Algal Res. **30**, 128–148 (2018)
2. Torres, M.D., Chenlo, F., Moreira, R.: Rheological effect of gelatinisation using different temperature-time conditions on potato starch dispersions: mechanical characterisation of the obtained gels. Food Bioprocess Technol. **11**, 132–140 (2018)
3. Liu, W., Yang, C., Zhou, C., Wen, Z., Dong, X.: An improved microwave-assisted extraction of anthocyanins from purple sweet potato in favor of subsequent comprehensive utilization of pomace. Food Bioprod. Process. **115**, 1–9 (2019)
4. Krystyjan, M., Ciesielski, W., Khachatryan, G., Sikora, M., Tomasik, P.: Structure, rheological, textural and thermal properties of potato starch - inulin gels. LWT - Food Sci. Technol. **60**, 131–136 (2015)
5. Fradinho, P., Sousa, I., Raymundo, A.: Functional and thermorheological properties of rice flour gels for gluten-free pasta applications. Int. J. Food Sci. Technol. **54**, 1109–1120 (2019)
6. Wu, M., Wang, J., Xiong, Y.L., Ge, Q., Yu, H.: Rheology and microstructure of myofibrillar protein–starch composite gels: comparison of native and modified starches. Int. J. Biolog. Macromol. **118**, 988–996 (2018)

7. Liu, C., Yan, X., Xu, X., Guo, B., Yang, R., Chen, J., Zhong, Y., Luo, S., Xu, J., Wu, J.: Changes in granular swelling and rheological properties of food crop starches modified by superheated steam. Starch/Staerke **71**, 1800132 (2019)
8. Torres, M.D., Fradinho, P., Rodríguez, P., Falqué, E., Santos, V., Domínguez, H.: Biorefinery concept for discarded potatoes. Int. J. Food Sci. Technol. (2019, submitted)
9. Bashir, K., Swer, T.L., Prakash, K.S., Aggarwal, M.: Physico-chemical and functional properties of gamma irradiated whole wheat flour and starch. LWT- Food Sci. Technol. **76**, 131–139 (2017)

Impact of Ca^{2+} on the Rheology of Hybrid Carrageenan from *Mastocarpus stellatus* and *Chondrus crispus* Red Seaweeds

M. D. Torres(✉), N. Flórez-Fernández, and H. Domínguez

Department of Chemical Engineering, Science Faculty, University of Vigo, Campus Ourense, Ourense, Spain
{matorres,noelia.florez,herminia}@uvigo.es

Abstract. Sol-gel diagrams and the corresponding rheological properties of kappa/iota-hybrid carrageenan hydrogels prepared with calcium counterions at different ionic strengths (<1 M) and biopolymer content (< 2%, w/w) were determined. For this purpose, two model red seaweeds such as *M.stellatus* and *C.crispus* were selected. Results indicated that the presence of calcium counterions allowed obtaining hydrogels with enhanced viscoelastic features. It should be remarked that sol-gel diagrams exhibited a wide range of gelled systems with different rheological properties attained varying the extracts source, biopolymer content and ionic strengths. This is critically relevant to expand the potential applications of the extracted biopolymers and meet the growing market demand of target biopolymers.

1 Introduction

Hybrid carrageenans are sulphated polysaccharides, which are conventionally isolated in hot alkali medium from red seaweeds belonging to the *Gigartinales*, *Rhodophyta* [1]. The three most commercially exploited carrageenans are kappa, iota and lambda carrageenans, which can be separately provided or as a well-defined mixture, since most of the seaweeds contain hybrid carrageenans [2, 3]. Overall, iota carrageenans form soft and elastic gels with higher gelling temperatures than kappa carrageenans [4]. Non-commercial kappa/iota-hybrid carrageenans extracted from *Mastocarpus stellatus* and *Chondrus crispus* red seaweeds could be promising alternatives to alleviate the natural polymers demand for innovative applications due to their high quality of hybrid carrageen [5]. These two model red seaweeds are geographically distributed in the Iberean Peninsula Coast (mainly Galicia and north Portugal) [6]. Hybrid carrageenans are not only demanded by their gelling ability, but also by their anticoagulant, antiviral, or immunomodulatory activities. These biopolymers showed promising potential to be developed as gelled therapeutic agents, although some controversy with pro-inflammatory properties arise [7].

Therefore, this work deals with the study of the phase diagrams (sol-gel transitions) of kappa/iota-hybrid carrageenan extracted from *M. stellatus* and *C. crispus* red seaweeds in the absence of alkali medium. The influence of calcium counterions on the

F. J. Galindo-Rosales et al. (Eds.): IBEREO 2019, SPM, pp. 79–82, 2020.
https://doi.org/10.1007/978-3-030-27701-7_17

rheological features of the prepared kappa/iota-hybrid hydrogels at different biopolymer content (<2%, w/w) and ionic strengths (<1 M) was discussed.

2 Materials and Methods

2.1 Raw Materials

Mastocarpus stellatus and *Chondru scrispus* red seaweeds (moisture content of 6.2 ± 0.3 and 5.9 ± 0.2 g/100 g, dry basis) were kindly provided by two Galician Companies, Porto-Muíños (Cerceda, Spain) and CEAMSA (Pontevedra, Spain), respectively. Seaweeds were stored in sealed plastic bags at room temperature and darkness until further processing to extract the hybrid carrageenan employed as raw material.

2.2 Hybrid Carrageenan Extraction

The hybrid carrageenan extraction was carried out following the conventional procedure previously reported for other carrageenophyte red seaweeds [8]. No alkali pretreatment was used during the extraction procedure. Briefly, milled seaweeds were mixed in distilled water for 2 h at 90 °C. After starch removal in the presence of α-amylase (50 °C for 1 h), supernatants were precipitated with ethanol (96%) and filtrated samples storage in the fridge until further use. Carrageenan extractions were made at least in duplicate.

2.3 Sol-Gel Diagrams

Phase diagrams for hybrid carrageenans extracted from both seaweeds were determined using a wide range of ionic (0.025, 0.05, 0.1, 0.25, 0.5, 0.75 and 1.0 mol/L) and biopolymer (0.5, 1.0, 1.5 and 2.0%, w/w) content. The corresponding aqueous hybrid carrageenan solutions heated up to 80 °C in the presence of Ca^{2+} were prepared in triplicate according to the procedure previously reported [2]. After preparation, all samples were stored at room temperature for 24 h. Visual and rheological characterisation of the samples was carefully performed in order to define the sol-gel diagrams.

2.4 Rheological Measurements

Mechanical spectra in terms of elastic modulus (G') and viscous modulus(G") versus angular frequency were conducted on a stress-controlled rheometer (MCR302, Anton PaarPhysica, Austria) using a plate-plate measuring system (1 mm gap, 25 mm diameter).

Viscoelastic behavior of aqueous solutions and gelled systems was determined within the linear viscoelastic region. Firstly, samples were placed on the rheometer-plate, edges sealed with light paraffin oil and were rested for 15 min. Then, stress sweeps were performed to define the linear viscoelastic region (<10 Pa for solutions and <20 Pa for gels). Mechanical spectra were conducted at room temperature (5 Pa for solutions and 10 Pa for gels). Rheological tests were performed in triplicate.

3 Results and Discussion

Figure 1 shows the sol-gel diagrams for hybrid carrageenan extracted from *M. stellatus* (left) and *C. crispus* (right), where the impact of the presence of Ca^{2+} on the sol-gel transition diagrams can be clearly observed.

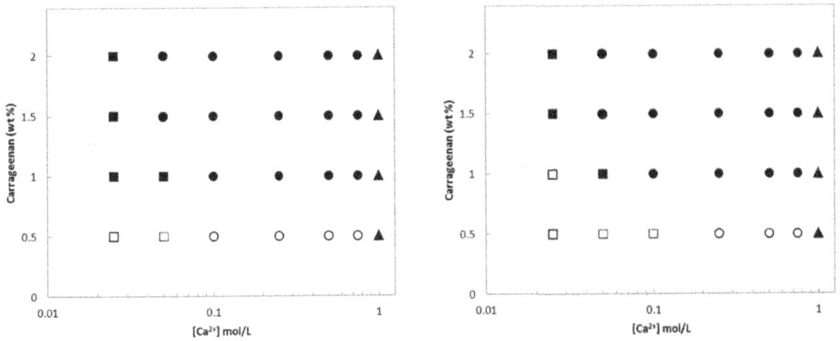

Fig. 1. Phase diagrams of hybrid carrageenan extracted from *M. stellatus* (left) and *C. crispus* (right) at different ionic content ($CaCl_2$). Symbols: clear solution (open squares), turbid solution (solid squares), clear gel (open circles), turbid gel (solid circles), triangle (stable suspension).

Preliminary sol-gel diagrams were performed by visual inspection with similar profiles. It can be noticed that hybrid carrageenan from *C. crispus* exhibited more clearer systems, although *M. stellatus* extracts featured clearer gels at lower ionic strength (0.1 mol/L). Clear gels involves an important advantage form the industrial point of view, since the presence of turbidity is not well accepted by the final consumer in wide range of food and non-food applications. In both cases, gels were achieved above 0.05 mol/L at the lowest biopolymer content (0.5%). In carrageenans from both seaweeds it was observed that systems with particle suspensions were reached above 0.075 mol/L ionic strengths. These results are consistent with those previously reported for hybrid carrageenan extracted from these red seaweeds in the presence of other smaller counterions (Na^+, K^+) [1, 2].

Figure 2 shows representative mechanical spectra of hybrid carrageenan from both seaweeds at the highest biopolymer content (2%) and intermediate ionic strength (0.5 mol/L). At these conditions, typical gel behavior with the elastic modulus higher than the viscous one and both moduli almost frequency independent was identified for hybrid carrageenan from both seaweeds. The magnitude of both moduli indicated that gels from *C. crispus* exhibited stronger (about one decade) mechanical properties than those formulated with *M. stellatus*. In both cases, intermediate strength gels were obtained. In all cases, the increase in ionic strength (up to 0.075 mol/L) and biopolymer content (2%) led to systems with stronger rheological properties. Gelled systems prepared with Ca^{2+} featured stronger properties than those reported for carrageenan from the same algae with smaller counterions [2].

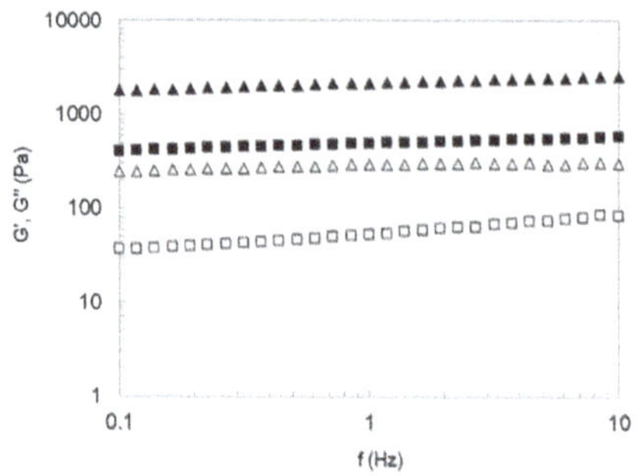

Fig. 2. Representative mechanical spectra of carrageenan from *M. stellatus* (squares) and *C. crispus* (triangles) red seaweed. Symbols: G' (elastic modulus, closed), G'' (viscous modulus, open).

4 Conclusions

The presence of calcium counterions allowed obtaining hybrid carrageenan gelled systems from *M. stellatus* and *C. crispus* with enhanced mechanical properties. The phase diagrams showed an interesting range of gelled systems with different mechanical properties resulting from the variation of biopolymer source and content as well as ionic strengths, with a broad range of potential applications.

Acknowledgements. Authors thanks Spanish Ministry of Science, Innovation and Universities (RTI2018-096376-B-100). M.D.T. thanks the same Ministry (IJCI-2016-27535), and N.F.F. thanks Xunta de Galicia (ED481B 2018/071) her postdoctoral grants.

References

1. Torres, M.D., Azevedo, G., Hilliou, L.: Phase diagrams of hybrid carrageenans extracted from *Ahnfeltiopsis devoniensis* and *Chondrus crispus*. Carbohydr. Polym. **136**, 449–458 (2016)
2. Azevedo, G., Torres, M.D., Sousa-Pinto, I., Hilliou, L.: Effect of pre-extraction alkali treatment on the chemical structure and gelling properties of extracted hybrid carrageenan from *Chondrus crispus* and *Ahnfeltiopsis devoniensis*. Food Hydrocoll. **50**, 150–158 (2015)
3. Campo, V.L., Kawano, D.F., Silva Jr., D.B., Carvalho, I.: Carrageenans: Biological properties, chemical modifications and structural analysis - a review. Carbohydr. Polym. **77**, 167–180 (2009)
4. Saha, D., Bhattacharya, S.: Hydrocolloids as thickening and gelling agents in food: a critical review. J. Food Sci. Technol. **47**, 587–590 (2010)
5. Ghanbarzadeh, M., Golmoradizadeh, A., Homaei, A.: Carrageenans and carrageenases: versatile polysaccharides and promising marine enzymes. Phytochem. Rev. **17**, 535–569 (2018)
6. van de Velde, F.: Structure & function of hybrid carrageenans. Food Hydrocoll. **22**, 727–734 (2008)
7. Pangestuti, R., Kim, S.J.: Biological activities of carrageenan. Adv. Food Nutrit. Res. **72**, 113–124 (2014)
8. Torres, M.D., Flórez-Fernández, N., Domínguez, H.: Impact of counterions on the thermo-rheological features of hybrid carrageenan systems isolated from red seaweed *Gigartinaskottsbergii*. Food Hydrocoll. **84**, 321–329 (2018)

Flow Behaviour of Vegetable Beverages to Replace Milk

Mariana Lopes, Carla Margarida Duarte(✉), Cristiana Nunes, Anabela Raymundo, and Isabel Sousa

LEAF - Linking Landscape, Environment, Agriculture and Food, Instituto Superior de Agronomia, Universidade de Lisboa, Tapada da Ajuda, 1349-017 Lisbon, Portugal

marianacl96@hotmail.com, {carladuarte,anabraymundo, isabelsousa}@isa.ulisboa.pt, crnunes@gmail.com

Abstract. Recently, milk consumption has been declining and there is a high demand for cow milk substitutes other than soy beverages. However, market offers are mainly cereal and nut-based beverages, which are essentially poor in protein content (less than 1.5% against the 3.5% in milk) and are not true milk replacers in that sense. Therefore, legume-based beverage was becoming a fast-growing segment in newer food development. This work presents the rheological behavior of commercial non-dairy beverages, showing that they are shear-thinning fluids and it is part of a database containing all information available about commercialized nondairy alternative beverages that is being collected to be used as the target for the development of pulse-based beverages. Flow properties of pulse-based beverages are expected to be typical non-Newtonian fluids, as current non-dairy alternative beverages, where the apparent viscosity decreases over shear, i.e., they are shear-thinning fluids which is fundamental for the mouth feel and consumer acceptance of new developing beverages. The viscosity curves obtained are presented as a guide flow pattern to be achieved by the use of rheology modifiers, if necessary, to find the right mouth feel.

1 Introduction

Although the milk segment is projected to account for the largest market share during the forecast period of 2018-2023, the market for dairy alternatives is projected to grow from USD 17.3 billion in 2018 to USD 29.6 billion by 2023, at a Compound Annual Growth Rate (CAGR) of 11.4%, and Asia-Pacific region represents the biggest market share [1, 2]. This growth is due to three main factors: (i) nutritional benefits offered by plant-based dairy alternatives, such as to reduce the cholesterol level, to improve cardiovascular health, and diabetes control [3]; (ii) the growing consumer preference for vegan diets; and (iii) increasing cases of lactose intolerance and milk allergies [4–6]. The resulting response from the Industry, so far, is the offer of cereal and nuts based beverages, such as those made from rice, almonds or oat that are lower in protein (<1.5%) than milk (3–4%) [7]. Legume-based beverages, such as those made from soy, can contain added vitamins and minerals and they contain a similar level of protein to

F. J. Galindo-Rosales et al. (Eds.): IBEREO 2019, SPM, pp. 83–87, 2020.
https://doi.org/10.1007/978-3-030-27701-7_18

milk (minimum 3%). Soy milk is still the most widely consumed non-dairy beverage, but has been decreasing its share because of health concerns related to GMO and allergens, high levels of isoflavones and with CO_2 footprint. The major issue that is hampering the production of legume-based beverage is the "beany" flavor, associated to endogenous lipoxygenases which oxidize unsaturated fatty acids in oil rich pulses like soy [8] and peanuts (over 20% fat), but should be less pronounced in pulses like peas, lupins or chickpeas (1.5 to 5% fat). The activity of the lipoxygenases is enhanced by the presence of polyphenols (bitter taste), but is suppressed by processing techniques like soaking, germination and pressure-cooking [9].

This work presents the rheological behavior of commercial non-dairy beverages and it is part of a database containing all information available about commercialized nondairy alternative beverages that is being collected for comparison to nutritional, rheological, physical and sensorial characteristics of the pulse-based beverages. This study aims to find a viscosity pattern to use as a guide to adjust the flow of legume-based beverages to be developed.

2 Materials and Methods

2.1 Material

Skimmed milk and eight different UHT non-dairy beverages were commercially acquired and stored at room temperature, until analysis: almond milk, oat milk, hazelnut milk, coconut milk, quinoa milk, nut milk, dry nuts milk and rice milk.

2.2 Rheological Measurements

Rheological measurements were performed using a controlled-stress rheometer (Haake MARS III, Germany), at 20 ± 1 °C. A serrated parallel plate, with 60 mm geometry, to avoid sample sliding was used, and the distance between plates was set at 0.25 mm, at 20 °C. The steady-state measurements were performed for viscosity curves determination, repeated three times for skimmed milk and each non-dairy beverage.

2.3 Statistical Analysis

Statistical analysis ANOVA One-Way, Means comparisons -Tukey Test, was performed using the software OriginPro version 8, to adjust flow curves to Power Law (Oswald-de-Waele) model to the shear-thinning region of the curves. The significance level was set at 95%.

3 Results and Discussion

3.1 Comparison of Skimmed Milk and Non-dairy Beverages

Non-dairy beverages evidenced typical non-Newtonian flow, where the apparent viscosity is decreasing over shear, i.e., they are shear-thinning fluids (Fig. 1).

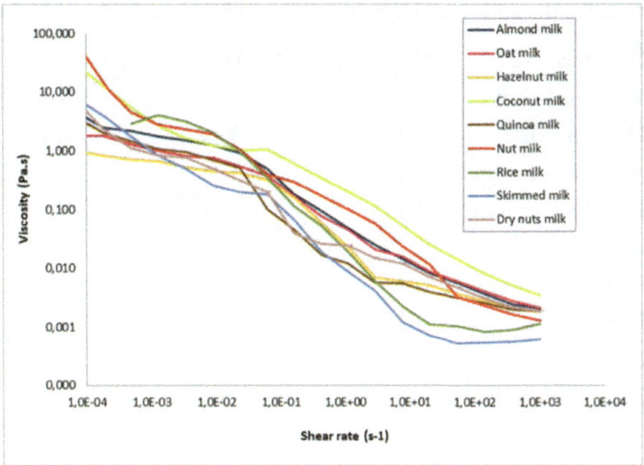

Fig. 1. Viscosity curves for the skimmed milk and the eight non-dairy beverages studied.

The skimmed milk and the beverages' consistencies were similar and varied between 0.014 and 0.065 Pa.s^n (Table 1). Skimmed milk and rice beverage were the ones with the lowest value of consistency. The coconut evidenced a higher consistency of 0.358 Pa.s^n which may be explained by the addition of rheology modifiers (higher amounts of gums added).

Table 1. Power Law model fitting results. (*corresponds to a significant different value at $p < 0.05$)

Samples	K (Pa.s^n)	n
Skimmed milk	0.014 ± 0.006	0.375 ± 0.050
Oat beverage	0.052 ± 0.021	0.246 ± 0.117
Almond beverage	0.065 ± 0.010	0.457 ± 0.049
Hazelnut beverage	0.047 ± 0.010	0.473 ± 0.031
Coconut beverage	0.358 ± 0.025*	0.262 ± 0.011
Quinoa beverage	0.026 ± 0.006	0.470 ± 0.060
Nut beverage	0.038 ± 0.005	0.592 ± 0.029
Dry nuts beverage	0.038 ± 0.004	0.439 ± 0.054
Rice beverage	0.008 ± 0.005	0.390 ± 0.031

As regards flow index, oat and coconut beverages presented the lower values, around 0.25. The skimmed milk and rice beverage showed "n" values around 0.38 and there is a third group of beverages which evidenced a higher flow index (0.439–0.592), not showing a correlation with its corresponding compositions.

From the resumed Table 2 of these beverages composition one can find that the protein content is from 0.1 to 0.8% (w/v), way far from the skimmed milk content of 3.5%.

Table 2. Skimmed milk and non-dairy beverages relevant composition (package labels).

Samples	Protein (% w/v)	Hydrocolloids	Sugars
Skimmed milk	3.5	No	No
Oat beverage	0.2	Yes	No
Almond beverage	0.5	Yes	No
Hazelnut beverage	0.4	Yes	Yes
Coconut beverage	0.1	Yes	No
Quinoa beverage	0.8	Yes	No
Nut beverage	0.2	Yes	No
Dry nuts beverage	0.6	Yes	Yes
Rice beverage	0.1	No	No

The lack of protein is compensated by the addition of hydrocolloids, and sometimes even sugar, to produce these beverages with acceptable mouth feel viscosity. This is a good example of the use of rheology modifiers and a bad example of nutritional replacement, that does not justify the higher market price of these vegetable beverages, when compared to milk.

4 Conclusions

Viscosity profiles of the vegetable beverages and skimmed milk were all shear thinning and showed extensive overlapping, which is an indication of the correct mouth feel target. This was attained by the use of rheology modifiers (added gums) to replace the protein effect of milk. The viscosity curves hereby presented will be used as a guide flow pattern to be achieved by the use of pulses to develop alternative beverages to milk, with similar protein content, to find the right mouth feel in the future legume-based beverages.

References

1. 28 February 2019. https://www.marketsandmarkets.com/Market-Reports/dairy-alternative-plant-milk-beverages-market-677.html
2. 19 March 2019. https://www.mintel.com/press-centre/food-and-drink/us-non-dairy-milk-sales-grow-61-over-the-last-five-years
3. Campos-Veja, R., Loarca-Piña, G., Oomah, B.D.: Review: minor components of pulses and their potential impact on human health. Food Res. Int. **43**, 461–482 (2010)
4. Lifschitz, C., Szajewska, H.: Cow's milk allergy: evidence-based diagnosis and management for the practitioner. Eur. J. Pediatr. **174**, 141–150 (2015)
5. Scrimshaw, N.S., Murray, E.B.: The acceptability of milk and milk products in populations with a high prevalence of lactose intolerance. Am. J. Clin. Nutr. **48**(4), 1142–1159 (1988)
6. 09 April 2019. https://www.transparencymarketresearch.com/plant-based-milk-market.html
7. Food Standards Australia, 28 February 2019. http://www.foodstandards.gov.au/consumer/nutrition/milkaltern/Pages/default.aspx

8. Yang, A., Smyth, H., Chaliha, M., James, A.: Sensory quality of soymilk and tofu from soybeans lacking lipoxygenases. Food Sci. Nutr. **4**(2), 207–215 (2016)
9. Khandelwal, S., Udipi, S.A., Ghugre, P.: Polyphenols and tannins in Indian pulses: effect of soaking, germination and pressure cooking. Food Res. Int. **43**(2), 526–530 (2010)

Polymers and Biopolymers

HNE Microemulsion: Development, Rheological Characterization and *In Vitro* Release Studies

Andreia Nunes[1(✉)], Joana Marto[2], Francisca Lopes[2], and Helena Margarida Ribeiro[2]

[1] Faculty of Pharmacy, Universidade de Lisboa, Lisbon, Portugal
andreia.a.nunes@campus.ul.pt
[2] Research Institute for Medicines (iMed.ULisboa), Faculty of Pharmacy, Universidade de Lisboa, Lisbon, Portugal
jmmarto@ff.ulisboa.pt, fclopes@ff.ul.pt, hribeiro@campus.ul.pt

Abstract. Human neutrophil elastase (HNE) is involved in the degradation of matrix proteins playing an important role in inflammation modulation. In this study, a microemulsion (ME) containing a new human neutrophil elastase inhibitor was developed and characterized in terms of pH, rheology (viscosity and oscillatory experiments) and drug release. This formulation is suitable for topical application to treat inflammatory diseases.

1 Introduction

Proteases are proteolytic enzymes that hydrolyses peptide bonds of proteins into smaller peptides or amino acids. In prokaryotic and eukaryotic cells, proteases control the cellular reactions through the cleavage of protein substrates. Serine proteases are the biggest known family of proteases, representing almost one-third of all proteases currently identified and are characterized by a serine residue in the active site of the enzyme. This family of proteases is known for presenting a catalytic triad composed of residues of aspartic acid (Asp), histidine (His) and serine (Ser), more specifically Asp102, His57 and Ser195.

Human neutrophil elastase (HNE) is a serine protease that belongs to the chymotrypsin superfamily and is stored and secreted from polymorphonuclear neutrophils.

HNE is involved in the degradation of matrix proteins such as elastin, collagen, fibronectin, playing an important role in the modulation of inflammation. Its action is controlled by endogenous inhibitors and when imbalanced it can lead to several pathological conditions, such as psoriasis, chronic obstructive pulmonary disease and acute respiratory distress syndrome. HNE inhibitors have the potential to be used as therapeutic tools for these diseases [1, 2].

Therefore, the synthesis and the delivery of new HNE inhibitors, able to modulate the proteolytic activity of HNE, represent promising therapeutics for all diseases where there is an excess in the production of this serine protease [1, 2].

F. J. Galindo-Rosales et al. (Eds.): IBEREO 2019, SPM, pp. 91–95, 2020.
https://doi.org/10.1007/978-3-030-27701-7_19

The aim of this work is the evaluation and characterization of HNE based microemulsion for skin delivery.

2 Materials and Methods

C21-C28 Alkane (Emosmart ™ C28), C15-19 Alkane (Emogreen ™ L15, Seppic ™) and a triglyceride-based oil more specifically Caprylic/Capric Triglycerides (Tegosoft® CT, Evonik). PEG-20 Glyceryl Triisostearate (Cithrol™ 10gtis, Croda Inc.) was used as surfactant, Propylene Glycol (Propylene Glycol EP from Mosselman) as humectant and Methyl and Propyl paraben (Nipagin™ M Sodium and Nipasol M Sodium™ from Clariant, respectively) as preservatives.

2.1 Microemulsion Preparation

The AAN9-based microemulsion was developed to mimic the skin lipid layer and for this purpose three different carbon chain oils were employed. The microemulsion was prepared using a cold method with the following components: water (5.97%), C21-C28 alkane (15.87%), C15-19 alkane (15.87%), caprylic/capric triglycerides (15.87%), PEG-20 glyceryl triisostearate (40.27%), propylene glycol (5.97%), methyl paraben (0.11%) and propyl paraben (0.01%). After the addition of all components the formulations were stirred with a glass rod.

2.2 pH

The pH values were determined twenty four hours after the production of the formulations and at room temperature (25 °C) using a potentiometer (METTLER TOLEDO).

2.3 Rheology

The rheological characteristics of the formulations were examined at high shear rates using continuous shear techniques and in the viscoelastic region using oscillation techniques. These experiments were performed with a controlled stress Kinexus Rheometer (Malvern) using a cone-plate geometry (truncated cone angle 4° and radius 40 mm). All the studies were performed at 25 °C. The viscosity method was carried out using a destructive measurement, where the shear stress of each formulation was obtained by increasing the shear rate from 0.1 s^{-1} to 100 s^{-1} and 8 samples per decade. In the oscillatory method, an amplitude sweep test was assessed where the shear strain varied between 0.01 Hz and 100 Hz, the frequency was 1 Hz and 10 samples per decade. Then, a frequency sweep test was performed with a shear strain of 5%, at frequencies between 0.01 Hz and 10 Hz and 10 samples per decade, at 25 °C.

2.4 Drug Release

In vitro drug release studies were performed using vertical Franz diffusion cells with Tuffryn®, 0.45 μm membranes. Water:ethanol (1:1, w/w) with 2.5% of PEG-40 hydrogenated castor oil were used as receptor phase. The study was performed for six hours at 37 °C and samples were collected at one hour intervals. The amount of released drug was analyzed by a fluorescence method and data was expressed in drug cumulative amount of permeated a function of time. An O/W emulsion and a solution with AAN-9 were used as controls.

3 Results and Discussion

3.1 pH and Rheology

The AAN9-based ME has a pH of 7.46, an expected value, in agreement with the pH of the components used. Figure 1 plots the flow curve for the AAN9-based ME exhibiting a shear thinning behavior at very low shear rates followed by a substantial Newtonian behavior. The shear-thinning behavior may be regarded as arising from modifications in the inner structure of ME.

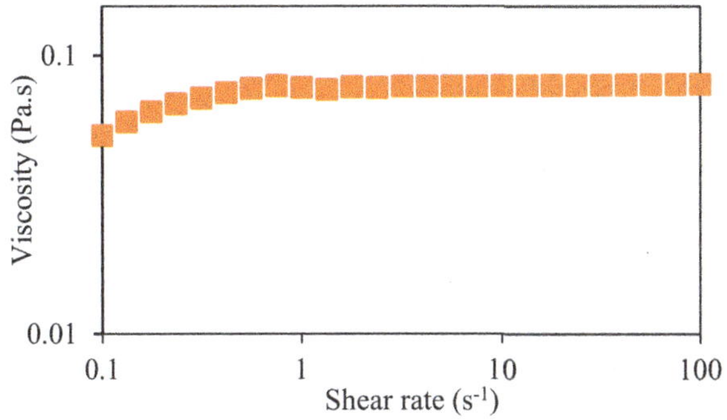

Fig. 1. AAN9-based ME: viscosity values Vs shear rate for the ME with AAN9 compound.

The oscillatory results are presented in Fig. 2, showing a pronounced value of the loss modulus, meaning a viscous prevalence over an elastic behavior. As expected, the loss tangent value at a frequency of 0.01 Hz is 12.92 indicating a liquid-like solution structure. This formulation has a weak 3D network structure with good skin spreadability. Nevertheless, the ME stability will not be affected because the stability results (data not shown) usually are a consequence of the ultralow interfacial tension between the oil and the water phases. The high oscillation values of the storage modulus are probably due to the experimental conditions that have been well selected for this ME. New rheological experiments should be carried out.

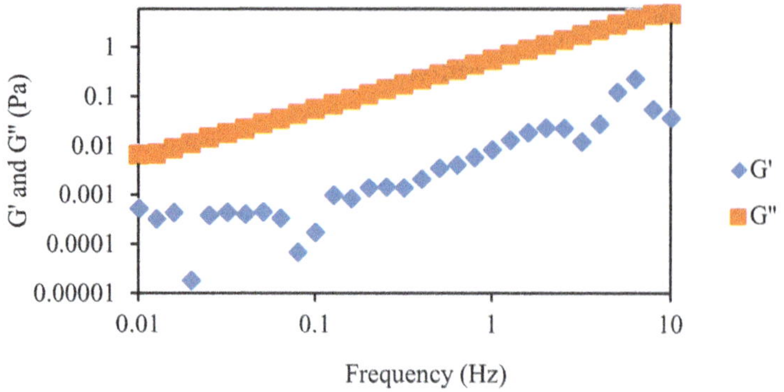

Fig. 2. AAN9-based ME: oscillatory results.

3.2 Drug Release

Topical delivery of some lipophilic and hydrophilic compounds appears to benefit from administrations of ME compared with conventional vehicles and rheology plays an important role because lower viscosities promote higher drug releases. In this study, the amount of AAN-9 released is strongly dependent on the type of formulation, as can be seen in Fig. 3. After six hours, the amount of AAN9 released was 2.90 ± 1.2 μg/cm^2 (10.0 ± 3.5%), 7.94 ± 2.1 μg/cm^2 (24.4 ± 7.4%) and 17.93 ± 1.6 μg/cm^2 (55.1 ± 5.7%) for ME, O/W emulsion and solution, respectively.

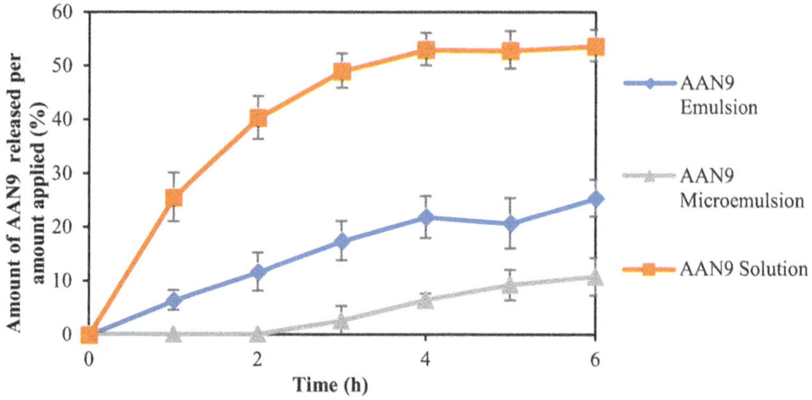

Fig. 3. Drug release studies for AAN9-based ME, emulsion with AAN9 and solution with AAN9 compound.

Regarding ME, a AAN9 controlled release would be expected, since this formulation contains a high percentage of surfactant and of an occlusive oil phase. The O/W emulsion has a surfactant that mimics the structure of the skin barrier and glycols which are known as cutaneous enhancers. The solution shows the highest percentage of AAN9 released over time. This liquid solution contained water: ethanol (1:1) with 2.5% of PEG-40 hydrogenated castor oil.

4 Conclusions

In spite that MEs have the advantage of being thermodynamically stable, the emulsion released four times more the amount of AAN-9. The results obtained were expected since skin diffusion is known to be dependent on structural properties of the 3D network of the vehicle and on the solubilization of the drug.

References

1. Marto, J., Ruivo, E., Lucas, S.D., Gonçalves, L.M., Simões, S., Gouveia, L.F., et al.: Starch nanocapsules containing a novel neutrophil elastase inhibitor with improved pharmaceutical performance. Eur. J. Pharm. **127**, 1–11 (2018)
2. Santana, A.B., Lucas, S.D., Gonçalves, L.M., Correia, H.F., Cardote, T.A.F., Guedes, R.C., et al.: N-Acyl and N-sulfonyloxazolidine-2, 4-diones are pseudo-irreversible inhibitors of serine proteases. Bioorg. Med. Chem. Lett. **22**, 3993–3997 (2012)

Rheology and Polymers: Born to Be Friends

Antxón Santamaría(✉)

POLYMAT and Polymer Science and Technology Department,
Faculty of Chemistry, University of the Basque Country (UPV/EHU),
Avda. Tolosa 72, 20018 San Sebastian, Spain
antxon.santamaria@ehu.eus

Abstract. The aim of this work is to present the two most important contributions of Rheology to the field of polymer science and technology: (a) Non Newtonian flow and relevance of shear thinning in polymer processing (b) Significance of chain entanglements in polymers viscoelasticity.

1 Introduction

In 1920, Hermann Staudinger, then professor of organic chemistry at the Eigenössische Technische Hochschule in Zurich, created a stir in the international chemical community when he postulated that materials, such as natural rubber, have very high molecular weights. In a paper entitled "Über Polymerisation" Staudinger presented several reactions that form high molecular weight molecules by linking together a large number of small molecules. During this reaction, which he called "polymerization," individual repeating units are joined together by covalent bonds.

The word "Rheology" was coined in 1929 on the occasion of the constitution of the Society of Rheology, under the auspices of Eugene C. Bingham and Markus Reiner, and was defined as the "Fundamental and practical knowledge concerning the deformation or flow of matter." The practical vocation of the new branch of science was clearly established from the very beginning: "The objects of this SOCIETY shall be the advancement of fundamental and practical knowledge concerning the deformation or flow of matter, hereafter designated as RHEOLOGY, and its connection with various other properties or applications of properties to industry…".

On behalf of the emerging disrupting industries, there was an enormous development of new materials in the first half of the 20th century. Polymers, in particular, started to play a crucial role in home appliances, medicine, cars, airplanes etc. The role of Rheology to characterize polymers and to establish processing conditions of these materials acquired a huge relevance. Rheological studies became, forever, an essential part of the knowledge of polymeric materials.

2 Reiner and Non-Newtonian Viscosity of Polymers

In October 1929 Markus Reiner published the paper entitled "Zur hydrodynamic von Systemen Veranderlicher Viscosität" where viscosity results of a rubber solution in benzene, calculated using a pressure driven rheometer, were reported. The lack of a

F. J. Galindo-Rosales et al. (Eds.): IBEREO 2019, SPM, pp. 96–99, 2020.
https://doi.org/10.1007/978-3-030-27701-7_20

linear correlation between the applied pressure and the obtained flow rate was observed. In particular, a ten times increase of pressure gave rise to a 13 times enhancement of the flow rate, indicating that the viscosity of the polymer solution was reduced with pressure, and consequently with shear rate, was increased. This was the first report reflecting a deviation from the linear dependence of the shear stress s_{21} on, shear rate $\dot{\gamma}_{21}$, $\sigma_{21} = \eta_0 \dot{\gamma}_{21}$ where η_0 is a constant called Newtonian viscosity. The results presented by Reiner in 1929 lead to the appearance of the viscosity function or non-Newtonian viscosity $\eta(\dot{\gamma})$. More in particular, the Non Newtonian viscosity of rubber/benzene solutions showed what is called a "shear thinning" behavior, or the viscosity reduction with shear rate. Besides of polymer solutions, polymer melts also show shear thinning as can be seen in Fig. 1 which shows the viscosity as a function of shear rate for two molten biodegradable polymers, polylactide, PLA, and polycaprolactone, PCL.

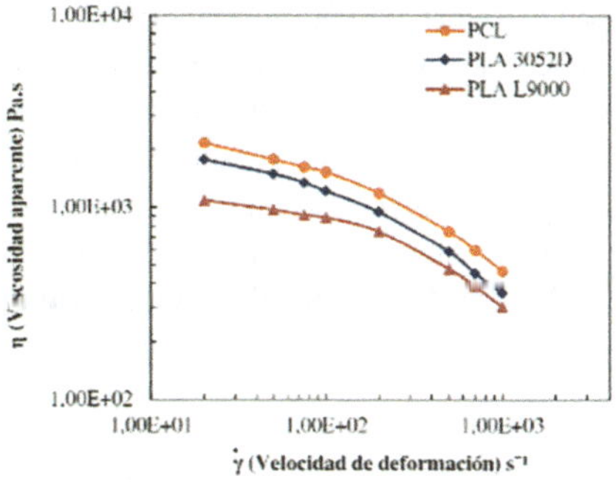

Fig. 1. Viscosity as a function of shear rates $\dot{\gamma}_{21}$ for the indicated polymers

The analysis of the viscosity function of molten polymers is very relevant for polymer processing. The shear rates involved in different polymer processing methods indicates that, for instance, in compression molding the shear rates are 1 to 10 s^{-1}, in 3D printing 100–1000 s^{-1} and injection molding >1000 s^{-1}. Needles to say, the viscosity decrease as shear rate is increased, favors processes which imply high shear rates, as is the case for injection molding. A singular example of the advantage of polymer melts being shear thinning is offered in Fig. 2, for the case of pressure driven 3D printing. The deviation from Newtonian behavior allows higher flow rates in the nozzle, which redounds in higher printing velocities.

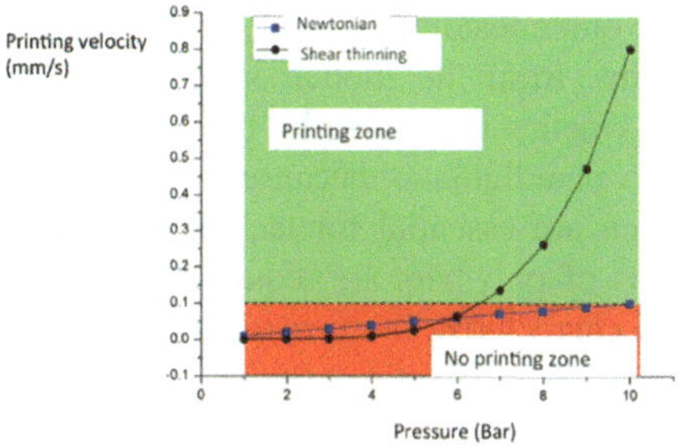

Fig. 2. Printing velocity as a function of pressure: advantage of shear thinning.

3 Polymer Chain Entanglements and Viscoelasticity

As the shear rate is decreased, the viscosity tends to a constant value or Newtonian viscosity, η_0, which can be obtained through adequate fitting of the experimental data (Fig. 1). The relatively high values of the viscosity found near the quiescent state are due to polymer chain interactions or entanglements, which, according to the model proposed by the 1991 Nobel Prize in Physics P. G. De Gennes, compel the chains to diffuse reptating in an entangled polymer based tube. Interestingly enough, reptation-tube model allows explaining the strong dependence of the Newtonian (low shear) viscosity on the molecular weight, M, of polymers: $\eta_0 = k\ M^{3.4}$. This equation, which is universal for any polymer melt, is widely used as a first approximation to the analysis of the length of polymer chains.

Moreover, polymer chain entanglements, that are kinetically present in the molten state, give rise to a physical network which imparts original and interesting viscoelastic properties to polymers. The relaxation modulus, $G(t) = \sigma_{21}(t)/\gamma_{21}$, IN THIS CASE THIS IS CORRECT which is a viscoelastic function that expresses how the stress gives up to an applied constant strain, reflects the presence of polymer chain entanglements in a unique way as compared to other materials. As can be deduced from Fig. 3, the presence of entanglements momentarily suppresses chains relaxation, creating what is called a "rubber plateau". At very low molecular weights (not shown in the figure) the entanglements plateau disappears, because the chains are too short to entangle and bring about a network.

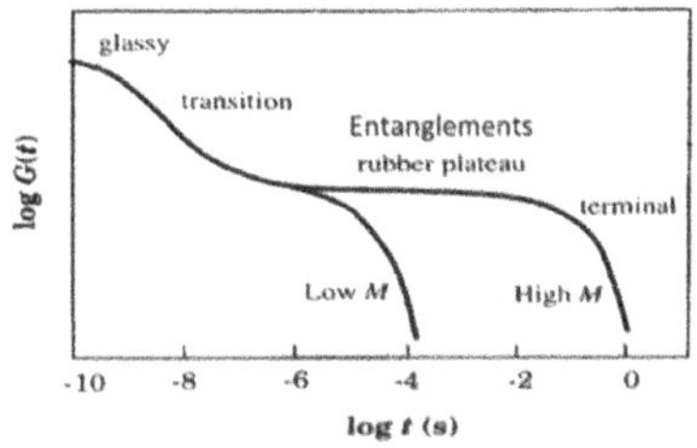

Fig. 3. The three viscoelastic zones of amorphous polymers

Entanglements in the molten state play an essential role in the final properties of the solidified polymers. Mechanical properties, such as tensile strength, elongation at break, impact factor and other features depend on the entanglement density reached when cooling the polymer from the molten state. In the case of semicrystallyne polymers, crystallization kinetics is affected by the entanglement density. The case of polymer adhesives is also paradigmatic, because polymer chain entanglements create temporary networks which are essential for tackiness or initial adhesion. Therefore, viscoelastic measurements of polymers are required to investigate the entanglements density and ultimately to characterize these materials.

4 Conclusions

Polymers are extraordinarily adequate materials for the development of rheological studies. Indeed, if there is any branch of science which owes practically everything to polymers, it is certainly Rheology. On the other hand, the advancement of polymeric materials would have been inconceivable with the crucial help of Rheology. Based on the relevance of the Non-Newtonian viscosity and the viscoelastic behaviour, Fig. 4 establishes the pivotal role of Rheology in all the steps of polymer science and technology.

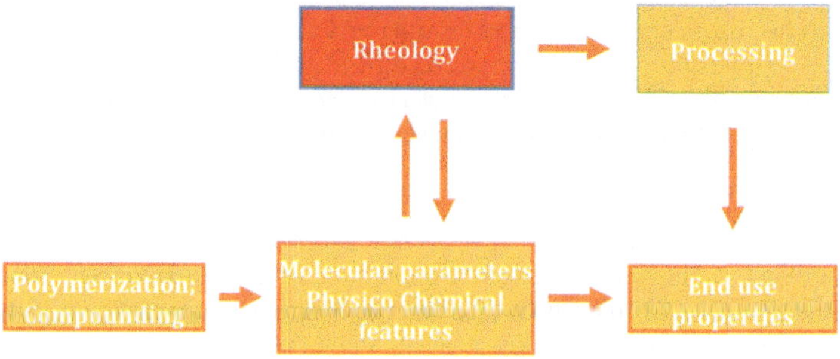

Fig. 4. Relationships between different aspects of the polymer science

Upgrading Recycled Polyolefins as Stimuli Responsive Materials

Leire Sangroniz[1], Ainara Sangroniz[1], Mercedes Fernández[1], Agustin Etxeberria[1], Alejandro J. Müller[1,2], and Antxón Santamaría[1(✉)]

[1] POLYMAT-Department of Polymer Science and Technology, Faculty of Chemistry, University of the Basque Country UPV/EHU, San Sebastian, Spain
{leire.sangroniz,ainara.sangroniz,mercedes.fernandez, agustin.etxeberria,alejandrojesus.muller, antxon.santamaria}@ehu.eus

[2] IKERBASQUE, Basque Foundation for Science, Bilbao, Spain

Abstract. Taking advantage of Joule heating effect, i.e. the increase of the temperature of the sample due to the application of an electric voltage, stimuli responsive materials have been prepared employing polypropylene and carbon nanotubes. The rheological measurements show that a percolated network has been formed, leading to an electrical percolation. As is typical for conductive materials, a significant increase of temperature is obtained when an electric current is applied. This preliminary study allows establishing the basis for the development of materials with applications based on the Joule heating effect.

1 Introduction

Stimulus responsive materials have attracted a great interest in the last years. Most of the studies are focused on the search of materials that can react to biological stimuli, in order to create new diagnosis techniques and drug delivery devices. On the other hand, there are also studies on materials that can change their shape under appropriate stimuli, like heat, electric or magnetic field, amongst others, which could have potential applications in biomedicine and also in engineering [1].

An effect that has been scarcely studied in polymer systems is the Joule heating effect, which can have interesting applications. In this work Joule heating of polyolefin based nanocomposites is studied, focusing also in the revalorization of recycled polyolefins. For that, semiconductor polymers are needed, which can be obtained dispersing conductive nanofillers in a polymer matrix [2, 3].

The aim of this work is to revalorize polyolefins waste, obtaining, in particular, nanocomposites based on recycled polypropylene and conductive nanofillers, such as carbon nanotubes.

Rheological measurements are a valuable tool to characterize recycled polymers, as well as polymer based nanocomposites, because, besides giving information on the molecular architecture of the chains, it also reports on the degree of dispersion of the filler. The rheological percolation is analysed and compared with the electrical

F. J. Galindo-Rosales et al. (Eds.): IBEREO 2019, SPM, pp. 100–103, 2020.
https://doi.org/10.1007/978-3-030-27701-7_21

percolation, allowing to select the materials potential ability to bring about Joule heating. The study of these systems will allow gaining insight on the stimulus responsive polymer nanocomposites with potential industrial applications.

2 Materials and Methods

2.1 Manufacture and Sampling

The polypropylene (PP) used in this work was supplied by Repsol, ISPLEN PP070G2M. Carbon nanotubes in the form of powder were from Cheap Tubes Inc. with a diameter of D = 30–50 nm and a length of L = 10–20 μm. The nanocomposite (PP/CNT) was prepared in a Collins co-rotating twin screw extruder at 210–230 °C and 150 rpm and, for comparison purposes, neat PP was prepared using the same procedure. The nanofiller content was 5% in weight. The samples were compression moulded at 180 °C to perform all the analysis.

2.2 Measurements

The rheological measurements were carried out employing an ARG2 TA Instrument rheometer using parallel plates and under nitrogen to avoid degradation at 180 °C. The electrical conductivity of the samples was measured at room temperature with an ARES rheometer, Dielectric Analysis option (DETS) that is coupled to a Novocontrol Interface. To analyse the Joule heating effect, i.e. the increase of the temperature of the sample due to the application of an electric voltage, an infrared FLIR camera was used with a home-made equipment.

3 Results and Discussion

Measurements employing small oscillatory shear were carried out in the linear regime. In Fig. 1, the elastic and loss moduli are plottedas a function of frequency for PP and PP/CNT. Polypropylene shows the typical behaviour of molten homopolymers, with the loss modulus higher than the elastic modulus at low frequencies. However, when CNTs are added solid like behaviour is observed and the elastic modulus becomes higher than the loss modulus. At the lowest frequencies G' shows values that are independent of frequency. This behaviour reflects that for the selected CNT content a percolated network has been formed due to the interactions between carbon nanotubes and polymer chains [2, 3].

If the dynamic viscosity is analysed, it can be observed that pure polypropylene shows a Newtonian viscosity of 3.4×10^3 Pa s at low frequencies and a significant reduction of the viscosity as the frequency increases. Nevertheless, PP/CNT nanocomposite shows a quite different behaviour. A viscoplastic behaviour is observed, characterized by an increase of the viscosity at low frequencies which results from the formation of a percolated network. Although at low frequencies the viscosity is significantly higher for the nanocomposite, in comparison with neat PP, at frequencies above 10 rad/s the differences are significantly reduced due to the pseudoplastic behaviour of the materials. Considering that frequencies and shear rates are equivalent, it can be stated that at high shear rates the presence of CNTs would not impair the processing of the material.

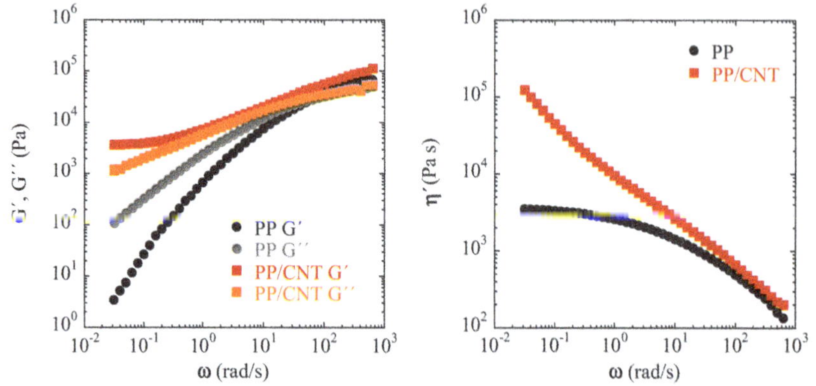

Fig. 1. On the left, elastic and loss moduli as a function of frequency for PP homopolymer and PP/CNT nanocomposite. On the right, dynamic viscosity as a function of frequency for PP and PP/CNT.

Electrical conductivity was measured at room temperature as a function of frequency and the results are displayed in Fig. 2. Polypropylene shows insulator behaviour characterized by a conductivity of 1×10^{-12} S/cm at low frequencies. For the PP/CNT nanocomposite the conductivity, reaches a value of 4.6 10^{-4} S/cm, which is independent of frequency indicating that the content of CNTs is above the percolation threshold.

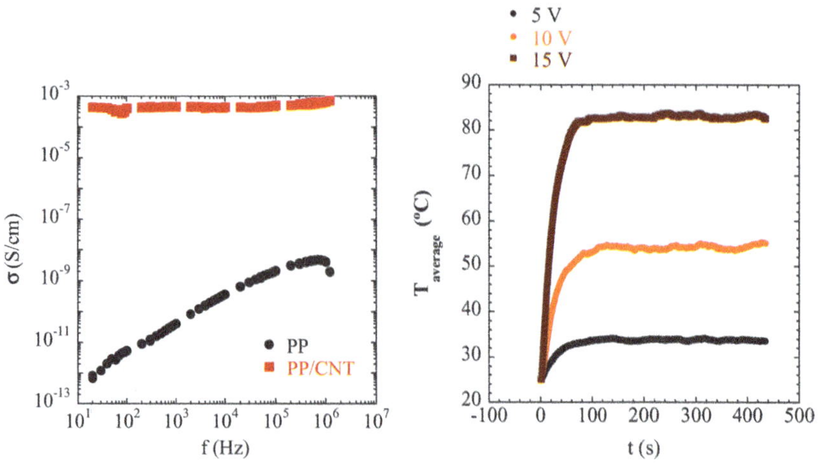

Fig. 2. On the left, conductivity against frequency for PP and PP/CNT measured at room temperature. On the right, the temperature increase of PP/CNT sample due to Joule effect at the indicated applied voltages.

In Fig. 2, the increase of temperature with time for various applied voltages are shown for a PP/CNT sample. As can be seen, after 100 s a constant value of the temperature is obtained. For the lowest voltage, 5 V, an increase in temperature of 8 °C is obtained, whereas for a voltage of 15 V the temperature rises up to 84 °C, with an increment of 60 °C. This temperature enhancement is high, as compared to the results reported in the literature.

4 Conclusions

In this work PP and a PP/CNT nanocomposite have been studied. The rheological are in agreement with conductivity results, showing a percolated network for PP/CNT nanocomposite. The formation of this percolated network allows using the polymer system to take advantage of the Joule heating effect, obtaining a considerable increase of the temperature for relatively low applied voltages. This study is crucial to the development of applications based on Joule heating effect.

References

1. Roy, D., Cambre, J.N., Sumerlin, B.S.: Future perspectives and recent advances in stimuli-responsive materials. Progr. Polym. Sci. **35**, 278–301 (2010)
2. Moniruzzaman, M., Winey, K.: Polymer nanocomposites containing carbon nanotubes. Macromolecules **39**, 5194–5205 (2006)
3. Pötschke, P., Abdel-Goad, M., Alig, I., Dudkin, S., Lellinger, D.: Rheological and dielectrical characterization of melt mixed polycarbonate-multiwalled carbon nanotube composites. Polymer **45**, 8863–8870 (2004)

Effect of Processing on the Rheological Properties and Water Uptake of Plasma Protein Superabsorbent Materials

Carlos Bengoechea(✉), Estefanía Álvarez-Castillo, José Manuel Aguilar, and Antonio Guerrero

Dpto. de Ingeniería Química, Escuela Politécnica Superior, Universidad de Sevilla, Seville, Spain
{cbengoechea, malvarez43, jmaguilar, aguerrero}@us.es

Abstract. As porcine plasma protein is mostly used as ingredient in foods (e.g. frankfurters) where water holding capacity is required, the feasibility of using it in the development of superabsorbent materials is considered in the present work. Plasma protein is blended with glycerol, which acts as plasticizer, at a 1:1 ratio, and subsequently injection-moulded at different mould temperatures, ranging from 60 to 120 °C. Moreover, the effect of pH is studied submitting the plasma protein to a pH modification procedure, involving aqueous dispersion, followed by pH modification and lyophilization. Following that procedure, bioplastics with pH equal to 3, 8 and 9 are produced. A reinforcement in the rheological properties was observed when plasma protein-based bioplastics are produced at higher moulding temperatures, resulting in a lower water uptake capacity. It is remarkable that at temperatures lower than 100 °C, superabsorbent materials are produced. Additionally, the highest water uptake is displayed by the sample prepared at pH 3, showing an apparent effect of pH on the properties of the samples. The estimation of the protein isoelectric point (IEP) is possible through zeta-potential and solubility tests performed on plasma protein aqueous dispersions at different pH values. The IEP is estimated to be around 5, which determines the positively charged character of the protein when processed at the acidic pH.

1 Introduction

Slaughterhouses currently produce an amount of blood much higher than that needed by the food industry, where it is used as water-holding agent in frankfurters, foaming agent, thickener or even as replacement of other ingredients, like eggs [1]. The surplus of blood should not be discarded directly to the environment due to its highly pollutant character [2]. Therefore, meat industry has been focused during the last decades to reduce its environmental impact by making a greater use of its raw materials and by-products [3]. In that sense, being blood an important source of proteins, the feasibility of producing biodegradable materials from plasma protein for applications such as packaging has been tested [4]. As the polar amino acids content in porcine plasma protein is relatively high, it may represent an adequate alternative to acrylic derivatives commonly used in the formulation of superabsorbent materials. Superabsorbents are

F. J. Galindo-Rosales et al. (Eds.): IBEREO 2019, SPM, pp. 104–107, 2020.
https://doi.org/10.1007/978-3-030-27701-7_22

materials that are able to absorb and hold an amount of water equal to more than 10 times their own initial weight, and those based on porcine plasma protein would benefit from being biodegradable. Processing temperature is considered as one of the most important parameters in the development of these materials, as it allows the adjustment of their swelling and absorption capacity. Thus, high temperatures generally promote the strengthening of the material's structure, finally hindering its water uptake. Moreover, pH has proven to strongly affect the protein behavior, as it may modify its conformation. Therefore, a modification in the pH may result in an alteration of the protein-water interactions, which eventually can lead to a different water uptake capacity.

The present manuscript focuses on the effect that both mould temperature and pH exert on the properties of porcine plasma based bioplastics processed through injection moulding.

2 Materials and Methods

2.1 Manufacture and Sampling

In the present work, a porcine plasma protein (PPP) concentrate, kindly provided by Essentia Proteins (Belgium), was used. Pharma grade glycerol (GL) from Panreac Química S.A (Spain) was used as plasticizer. To modify the pH, flour was dispersed in deionized water and monoprotic acid (HCl) or basis (NaOH) were used to reach the desired pH (3.0 and 10.0). Then, the solution was frozen and freeze dried.

PPP and GL were mixed in a constant ratio of 1:1 in a mixer rheometer Polylab QC (ThermoHaake, Germany) at room temperature during 5 min. The homogeneous doughs were injected in a Minijet Piston (ThermoHaake, Germany) during 150 s. The cylinder of the injector was set at 40 °C and the mould was set at different temperatures in order to find the optimal value to obtain superabsorbent properties.

2.2 Dynamic Mechanical Thermal Analysis

Oscillatory measurements were carried out at small strain amplitude in torsion mode in order to characterize the behavior of PPP rectangular bioplastic samples as the temperature increased from −30 to 140 °C (5 °C/min) in a DHR-3 hybrid rheometer (TA Instruments, USA) with a rectangular geometry. The frequency was kept constant at 1 Hz, and the strain was selected in each case to ensure the linear viscoelastic region (LVR) was not overpassed.

2.3 Water Uptake Capacity

That test consisted of a 3-steps procedure: in a first moment, samples were placed in an oven at 50 °C until constant weigh (w_1); then samples were immersed in deionized water during 24 h and weighed (w_2); and finally, they were placed back in the oven until a new constant weigh was reached (w_3). increase between w_3 and w_2 represent the amount of water absorbed by the material.

3 Results and Discussion

Probes injected at different mould temperature were submitted to temperature sweep essays in torsion mode. Storage (G') and loss (G") moduli profiles are shown in Fig. 1 for samples at pH 8 (native pH).

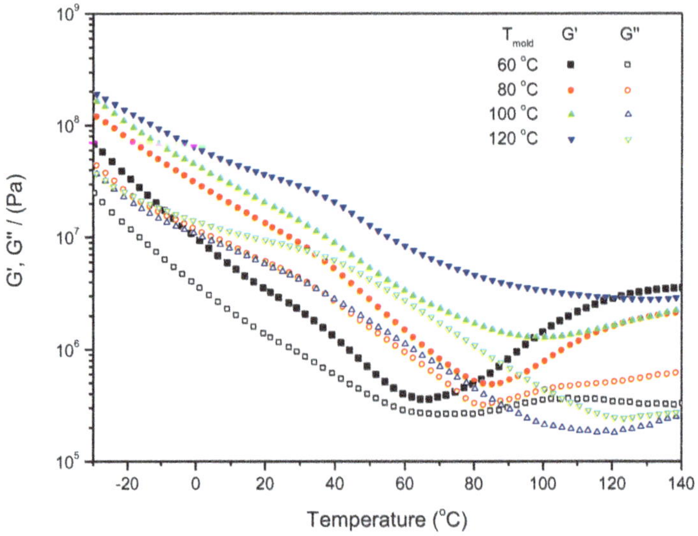

Fig. 1. Storage and loss moduli for rectangular samples of PPP/Gl obtained by injection at different mould temperatures.

It can be observed that the minimum G' was reached at mould temperature employed, and increasing with mould temperature was increased. Thus samples injected at mild temperatures shown a higher thermosetting potential, while samples injected at higher temperature show initially higher values in G' and G". Thus, it is confirmed that strengthening is promoted at higher temperatures.

Moreover, it should be highlighted that superabsorbent materials were only obtained when samples were processed at 60 and 80 °C, with water uptake values around 2000 and 1250%. At higher temperatures, water uptake was always equal or lower than 500% (Table 1). This may be explained on basis of the strengthening that is promoted by temperature that reduces the swelling ability of the sample and, therefore, its water uptake. Most thermal induced cross-linking reactions taking place in proteins are associated to the formation of sulfide bonds.

When observing the effect of pH on water uptake in Table 1, it may be observed how all materials processed at the mildest temperature (60 °C) displayed a superabsorbent character, regardless of pH. Notwithstanding, absorption capacity was remarkably higher at acidic pH, which may be explained on basis of a higher deformability shown in tensile tests.

Table 1. Water uptake capacities for samples ppp/Gl obtained at different mould temperatures or different pH.

		Water uptake capacity (%)			
		Mould temperature (°C)			
		60	80	100	120
pH	3	3726 ± 335			
	8	1942 ± 499	1314 ± 133	496 ± 113	209 ± 15
	10	1120 ± 133			

4 Conclusions

From the results presented it could be concluded that superabsorbent materials could definitely be obtained from porcine plasma protein through a careful control of the process parameters, without any chemical modifications. Furthermore, it was possible to increase the superabsorbent capacity if the pH of the concentrate is lowered to acidic pH values. If mould temperature was higher than 80 °C, the reinforcement produced in the material hinders its swelling capacity and the superabsorbent threshold is not overpassed.

Acknowledgements. Authors acknowledge support provided by Ministerio de economía y competitividad and FEDER (CTQ2015-71164-P and RTI2018-097 100-B-C21).

References

1. Hurtado, S., Saguer, E., Toldrà, M., Parés, D., Carretero, C.J.: Meat Sci. **90**(3), 624–628 (2012)
2. Álvarez, C., Rendueles, M., Díaz, M.: J. Agric. Food Chem. **60**(22), 5636–5643 (2012)
3. Otles, S., Despoudi, S., Bucatariu, C., Kartal, C.: Food Waste Recover, pp. 3–23 (2015)
4. Nuthong, P., Benjakul, S., Prodpran, T.J.: Food Sci. Technol. **42**(9), 1545–1552 (2009)

S. aureus and *E. coli* Co-culture Growth Under Shear

Raquel Portela[1], Pedro L. Almeida[2,3], Rita G. Sobral[1], and Catarina R. Leal[2,3](✉)

[1] UCIBIO@REQUIMTE, Faculdade de Ciências e Tecnologia, Universidade Nova de Lisboa, 2829-516 Caparica, Portugal
rp.portela@campus.fct.unl.pt, rgs@fct.unl.pt
[2] Área Departamental de Física, ISEL – Instituto Superior de Engenharia de Lisboa, Rua Conselheiro Emídio Navarro 1, 1959-007 Lisbon, Portugal
{palmeida, cleal}@adf.isel.pt
[3] CENIMAT/I3N, Departamento de Ciência dos Materiais, Faculdade Ciências e Tecnologia, Universidade Nova de Lisboa, 2829-516 Caparica, Portugal

Abstract. Growing monocultures of two different species of human commensal/pathogenic bacteria, *Staphylococcus aureus* – a non-motile gram-positive coccus and *Escherichia coli* – a motile gram-negative rod, were characterized using a real-time *in situ* rheology and rheo-imaging strategy. Subjecting bacterial populations to a shear flow is a closer approximation to bacterial thriving in the host, where they experience mechanical forces such as arterial or venous pressure. For both cultures, as the cell density of the population increases, cells rearrange themselves in different aggregates, capable of strongly influencing their environment, and leading to very different physical rheological responses, where motility appears to be determinant. One of the most striking observations is the behavior of the viscosity growth curve, showing dramatic value variations, with no counterpart in traditional biological measurements, as well as the coupling between translational and rotational motion of the *E. coli* aggregates along the growth curve [1], while *S. aureus* cells tend to sediment [2], over long periods of time. In the present study, a similar approach was applied to a co-culture of these two bacteria, *S. aureus* and *E. coli*, to evaluate the effect of possible interspecies interactions on the viscosity curve of the culture, during growth, when subject to a shear flow. Surprisingly, the observed behavior of the viscosity growth curve was enhanced in comparison to each individual curve and reveals a combination of details specific of each monoculture, suggesting synergy between these two bacterial species. After the rheological analysis, the final co-culture was recovered and inoculated on different solid media that allow to distinguish the development of *S. aureus* or *E. coli* colonies. Unexpectedly, *S. aureus* showed the capacity to accelerate its growth rate relatively to *E. coli*, when the two-species community is subjected to a shear flow. This behavior may reflect the occurrence of specific growth adaptations during co-culture upon shear flow, getting one step closer to physiological conditions.

F. J. Galindo-Rosales et al. (Eds.): IBEREO 2019, SPM, pp. 108–112, 2020.
https://doi.org/10.1007/978-3-030-27701-7_23

1 Introduction

In most ecological niches, including the human host, microbes frequently co-exist in complex mixed communities, rather than as single species populations. Microbe-microbe interactions can be of various types, ranging from antagonistic relations in which the different species compete for the same niche, to mutualistic scenarios, in which the different species all profit. The type of interaction will determine the overall composition and dynamics of a given microbial community [3]. In fact, the co-existence of different microorganisms is dependent on numerous parameters, such as the environmental preferences, the nutritional requirements or the quorum sensing mechanisms of each species, to name a few. Such varying parameters will result in a high diversity of species interaction patterns as described in [4].

In the specific scenario of bacterial biofilms, these are frequently found to be constituted by dual or multiple species, instead of a single species. Cell-to-cell interactions impact the temporal and spatial organization of biofilm design and, as for other mixed population settings, may be classified as either cooperative or competitive. In this regard, the study of a co-culture system is an appropriate approach to address how environmental changes may affect species interactions, such as their relative abundance, distribution and dynamics of development.

In this communication, we chose to study a dual species system composed of two human bacteria, that can both colonize the host as commensal organisms, or can be pathogenic, depending on the strain-specific virulence factors and on host-associated conditions. In particular, we decided to determine how an imposed shear stress would affect the viscosity profile of a dual-species population during growth, compared to the individual profiles of each of the bacterial species grown as a single-species population.

2 Experimental Procedures

2.1 Bacterial Cultures and Growth Conditions

In this study two different bacterial species were used: the methicillin resistant *Staphylococcus aureus* (MRSA) strain COL, a gram-positive human pathogen of nosocomial origin [5] and *Escherichia coli* DH5α strain (Invitrogen), a gram-negative engineered strain routinely used in laboratory for genetic engineering purposes. The strains were grown overnight in tryptic soy agar medium (TSB agar) plates at 37 °C. On the next day a single colony of each strain was inoculated in 5 mL of fresh tryptic soy broth (TSB) and grown overnight at 37 °C in an orbital shaker at 180 rpm to promote the aeration of the cultures.

The monocultures of *S. aureus* and *E. coli* were set by inoculating fresh TSB medium with the starting cultures to obtain an initial OD620nm = 0.005.

The co-culture of the two bacterial species, *S. aureus* and *E. coli*, was set by co-inoculating fresh TSB medium with individual starting cultures at a 1:1 ratio, in order to obtain an initial OD620nm = 0.005. The individual cultures and the co-culture were incubated at 37 °C with aeration in an orbital shaker (180 rpm). The growth curves were determined by monitoring the OD620 nm along time for 600 min, at discrete time points, using a spectrophotometer (Ultrospec 2100 pro).

2.2 Characterization Techniques

Optical microscopy

A Leica DMR microscope associated to a Leica DFC320 camera and Leica IM500 Image software V1.20, were used to characterize the aggregation processes during growth. For each aliquot, at specific growth moments, 120, 300, 420 and 600 min, a calibrated volume of sample of 10 μL was observed and at least 10 photos were randomly taken. From these images the bacteria distribution during co-culture growth was evaluated, namely the number of clusters of each bacteria strain vs time.

Rheology

A controlled stress rotational rheometer Bohlin Gemini HRnano equipped with a steel plate/plate geometry, with diameter 40 mm and 2000 μm gap (to ensure a good signal), was used to measure the viscosity growth curve at a constant shear rate of 10 1/s (which mimics the incubator conditions), during 900 min over the same culture sample. Measurements were performed at 37 °C to allow optimal bacterial growth. The viscosity of TSB culture medium was also measured in identical conditions and presents an average value of 0.41 ± 0.17 mPa.s.

3 Results and Discussion

Representative curves of the steady-state shear viscosity growth curve for the *S. aureus* and *E. coli* co-culture are presented in Fig. 1.

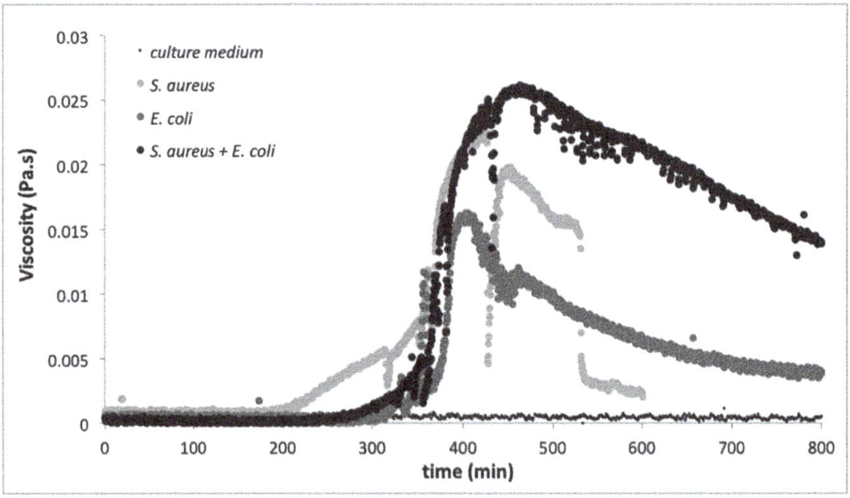

Fig. 1. Viscosity growth curve measured at constant shear rate of 10 1/s for the individual bacterial cultures *S. aureus* (light grey) and *E. coli* (dark grey) and for the bacterial co-culture *S. aureus* + *E. coli* (black) (representative curves). All measurements were performed at 37 °C.

An increase of ~70x, with respect to the medium culture viscosity value was observed to occur during the *exponential phase* at ~450 min. After this maximum value, the viscosity growth curve followed a monotonous behavior with a decreasing rate slower than the observed in the results obtained for each individual bacterium culture. This result showed that a complex structure is developed during the co-culture growth process, indicating that a synergistic cooperation between these two bacterial species may be occurring. These results are also supported by the optical densities values and the microscopy analysis presented in Fig. 2.

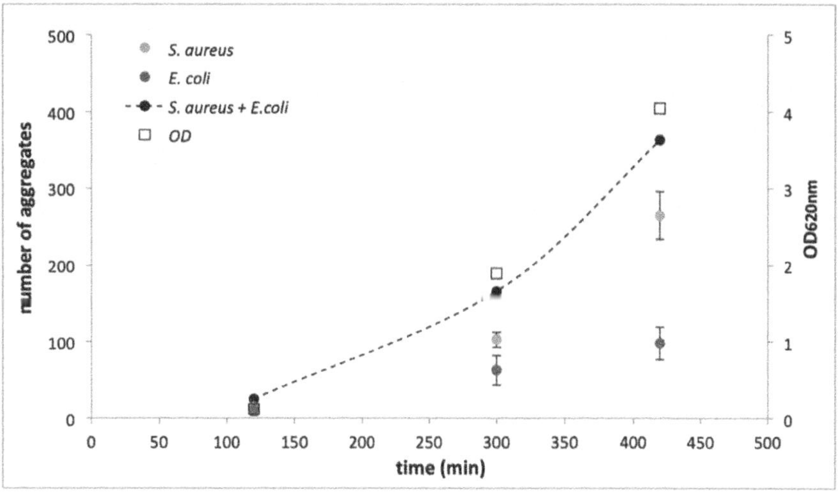

Fig. 2. Number of aggregates during the co-culture *S. aureus* + *E. coli* (black) growth: of each bacteria strain, *S. aureus* (light grey) and *E. coli* (dark grey); and respective optical densities, OD620 nm. Dashed line is a guide for the eye. All measurements were performed at 37 °C. Error bars represent the standard deviation of the mean.

Comparable results regarding the characterization of mixed species interactions were obtained with mixed dual-species biofilm growth where several gram-negative species were considered as well as *S. aureus* [6].

Acknowledgments. Strain COL was a kind gift from H. de Lencastre and A. Tomasz. This work was supported by FEDER through COMPETE 2020; FCT Projects No. UID/CTM/50025/2013, PTDC/FIS-NAN/0117/2014 (P.L.A.) and PTDC/BIA-MIC/31645/2017 (R.G.S.); and UCIBIO, which is financed by national funds from FCT/MEC (Grant No. UID/Multi/04378/2019). SPR also contributed financially to the presentation of this work.

References

1. Portela, R., et al.: Phys. Rev. E **94**, 062402 (2016)
2. Portela, R., et al.: Eur. Phys. J. E **42**, 26 (2019)
3. Mitri, S., Richard, F.K.: Annu. Rev. Genet. Annu. Rev. **47**, 247 (2013)
4. Das, P., et al.: PLoS ONE **13**(3), e0195161 (2018)
5. Gill, S.R., et al.: J. Bacteriol. **187**, 2426 (2005)
6. Makovcova, J., et al.: Microb. Biotechnol. **10**(4), 819 (2017)

Synergic Effects on the Viscosity of Sodium Carboxymethylcellulose in Mixtures with Xanthan, Guar and Tara Gum

Fabian Ramos[1(✉)] and Isabel Hernández[2]

[1] Materials and Nanotechnology Area, Facultad de Minas, Universidad Nacional de Colombia, Campus Robledo, 050034 Medellin, Colombia
framosmal@unal.edu.co
[2] Chemical and Petroleum Engineering Area, Facultad de Minas, Universidad Nacional de Colombia, Campus Robledo, 050034 Medellin, Colombia
maihernandezmo@unal.edu.co

Abstract. Sodium carboxymethylcellulose (CMC), Guar Gum, Xanthan Gum and Tara Gum are hydrocolloids used in the food industry as stabilizers and thickeners which individually have interesting properties in applications. When these products are physically mixed, they generate an effect know as synergy, in which the rheological properties of the hydrocolloids are not proportional to the composition of the mixture, but are enhanced. This phenomenon is due to the intermolecular interactions of materials in dissolution [1]. Synergistic properties between the hydrocolloid mixtures are of great interest because new products can be made with high viscosities, better performances and economic benefits in food applications.

1 Introduction

Several gums are synergistic with other gums, meaning that the net viscosity or gel strength, when the two gums are used together, is greater than would be expected from the additive combinations of each gum [2]. The gums form an interaction that creates a more effective three-dimensional network to trap water.

Figure 1 illustrates the differences between chemical structural of gums. The synergistic effects between the gums are given by the amount of hydroxyl ions present in each of the polymer chains that form the gums and how the networks are formed through hydrogen bonds in the water.

Therefore, the aim of this research is to describe the viscosity behavior of Guar Gum, Xanthan Gum and Tara Gum in pure components, binary and ternary mixtures with sodium carboxymethylcellulose in neutral aqueous solution.

F. J. Galindo-Rosales et al. (Eds.): IBEREO 2019, SPM, pp. 113–117, 2020.
https://doi.org/10.1007/978-3-030-27701-7_24

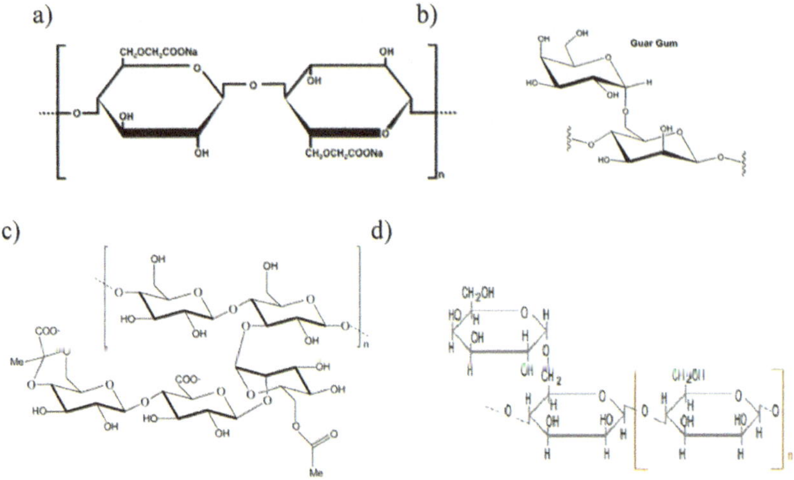

Fig. 1. Chemical structure of (a) CMC, (b) Guar Gum, (c) Xanthan Gum and (d) Tara Gum.

2 Materials and Methods

Viscosity tests are made in neutral solutions (pH = 7.0). Guar Gum, Xanthan Gum, Tara Gum and Sodium Carboxymethylcellulose (CMC) were individually weighed in different proportions, making the correction by weight according to the moisture content of each product (Dry Basis Viscosity).

Powders are mixed physically and added to water simultaneously. Viscosity of the mixtures were made at 1% wt. concentration to the solution. Measurements were taken when the product is dissolved at a temperature of 25 °C. Readings are made using Brookfield RVT viscometer, spindle 4 at 20 rpm.

For the determination of the viscosity points, a {3,5} simplex-lattice design of experiments with mixtures was chosen to obtain 21 points of viscosity. The ternary composition diagrams and the visualization of the behavior of the viscosity with respect to the composition was made using Matlab® (Table 1).

Table 1. Moisture and viscosity of the gums

Product	Moisture (%)	Brookfield viscosity 1%, 20 rpm (cP)
CMC	6.11	6200
Xanthan Gum	13.92	3300
Guar Gum	11.83	5300
Tara Gum	10.89	6000

3 Results and Discussion

Figures 2, 3 and 4 illustrate the viscosity behaviors for different composition points of the mixtures between the gums and CMC. Color bar scale situated to right of each composition triangle shows the color corresponding to the viscosity. Dark blue regions have the lowest viscosity, while dark red regions correspond to highest viscosity.

In CMC-Guar-Tara system (Fig. 2), higher viscosities and synergies are presented in the ternary mixture when the compositions are maintained of 20 to 60%, or binary mixtures between CMC-Guar gum or CMC-Tara gum.

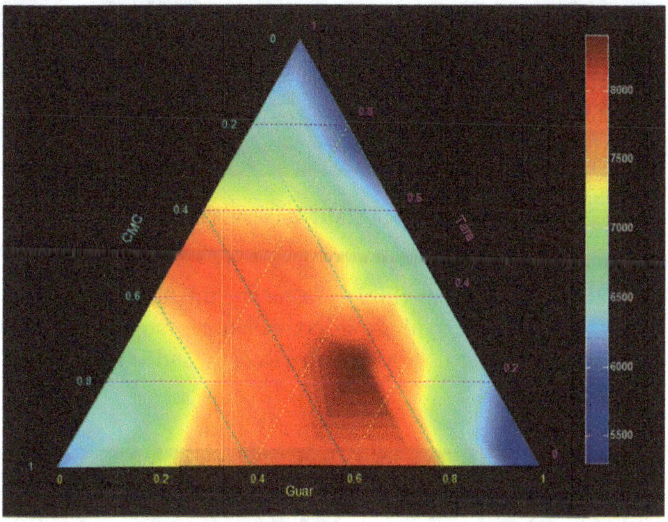

Fig. 2. Viscosity of CMC-Guar-Tara system

In CMC-Guar-Xanthan system (Fig. 3), the highest synergistic effect is found for 75% of guar gum and 25% of xanthan gum. By adding CMC to a mixture of guar gum and xanthan the synergy of the two gums is diminished.

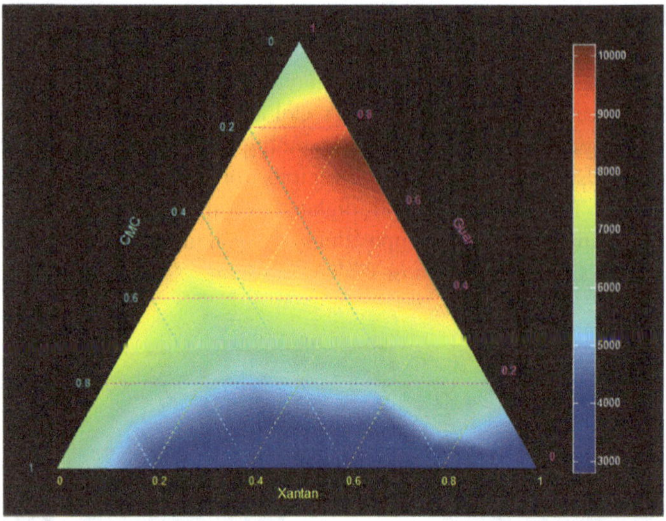

Fig. 3. Viscosity of CMC-Xanthan-Guar system

In CMC-Xanthan-Tara system (Fig. 4), a binary mixture between Xanthan Gum and Tara Gum generates high synergy in compositions ranging from 20% to 80%. Adding CMC to this system causes a drop in viscosity and synergy.

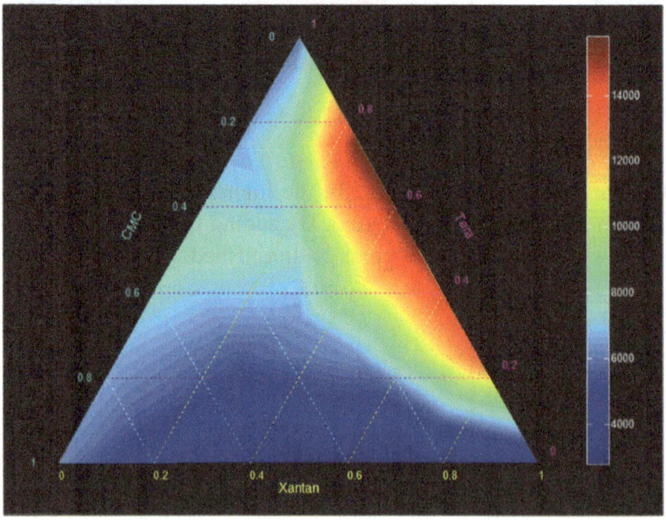

Fig. 4. Viscosity of CMC-Xanthan-Tara system

4 Conclusions

The viscosity behaviors of Guar, Xanthan and Tara gums as pure components and binary and ternary mixtures with CMC in neutral aqueous solution show maximum and minimum viscosities due to synergistic effects between the interactions of the materials.

The CMC-Guar-Tara system was the only system where adding CMC represents an advantage from the point of view of the viscosity and synergy effects in a wide range of compositions. In contrast, in the CMC-Xanthan-Tara system, it is counterproductive to add CMC because the viscosity is decreased.

References

1. Tipvarakarnkoon, T., Senge, B.: Rheological behavior of gum solutions and their interactions after mixing. Annu. Trans. Nordic Rheol. Soc. **16**, 1–8 (2008)
2. Laaman, T.: Hydrocolloids in Food Processing. IFT Press, Wiley-Blackwell, Chicago (2011)
3. Ospina, M., Sepulveda, J., Restrepo, D., Cabrera, K., Suarez, H.: Influencia de La goma xantan y goma guar sobre las propiedades reológicas de leche saborizada con cocoa. Revista biotecnología en el sector agropecuario y agroindustrial **10**(1), 51–59 (2012)
4. Weber, F., Clerici, M., Collares Queiroz, F., Chang, Y.: Interaction of guar and xanthan gums with starch in the gels obtained from normal, waxy and high-amylose corn starches. Starch J **61**, 28–34 (2009)
5. Westra, J.: Rheology of (carboxymethyl)cellulose with xanthan gum properties. Macromolecules **22**, 367–370 (1989)
6. Whistler, R., Bemiller, J.: Industrial Gums, Polysaccharides and Their Derivatives, 3rd edn. Academic Press, San Diego (1993)
7. Wu, Y., Ding, W., Jia, L., He, Q.: The rheological properties of tara gum (caesalpinia spinosa). Food Chem. **168**, 366–371 (2015)

Understanding the Diffusion and Rheology of Unentangled Associating Polymers with Simulations

Mahmoud Bagheri, Andrés R. Tejedor, and Jorge Ramírez(✉)

Department of Chemical Engineering, Universidad Politécnica de Madrid, Madrid, Spain
jorge.ramirez@upm.es

Abstract. In this work, we briefly review a recently developed model to study the diffusion and rheology of associating polymers. The model describes successfully the experimental data, including an unexpected phenomenological superdiffusive scaling at distances which are larger than the molecular size. The understanding of the dynamics of associating polymers can have a big impact in applications such as biocompatible gels or self-healing materials.

1 Introduction

Associating polymers are a class of macromolecules that contain chemical groups (*stickers*) that can create intermolecular reversible bonds mostly by means of non-covalent interactions like H-bonds, metal-ligand coordination, ionic or hydrophobic/hydrophilic interactions [1]. The energies of these bonds are of the order of a few $k_B T$, where k_B is Boltzmann's constant and T is the temperature, so they can associate and dissociate easily at room temperature, forming viscoelastic soft materials. These materials have many potential technological applications as, for example, sacrificial components in tough physical double networks [2], as synthetic matrices for tissue engineering [3], as injectable biomaterials for minimally invasive surgery [4] or as self-healing soft materials [5]. In all these applications, it is crucial to understand and predict the dynamical response of the material, in the form of its mechanical properties or the diffusion of the network forming macromolecules.

Recent forced Rayleigh scattering (FRS) experiments made by the Olsen group at MIT have showcased how a variety of unentangled associative polymers with different architecture and different nature of the associating interactions exhibit an unexpected and rich diffusion behaviour [6, 7], including a phenomenological superdiffusive regime at lengthscales that are larger than the molecular size, which crosses over to a Fickian regime at lengths that are several orders of magnitude larger than the radius of gyration of the molecules that form the transient network. This experimental evidence was unexpected, because in principle all diffusion should be Fickian at length scales larger than the molecular size.

F. J. Galindo-Rosales et al. (Eds.): IBEREO 2019, SPM, pp. 118–122, 2020.
https://doi.org/10.1007/978-3-030-27701-7_25

2 Model

Trying to understand these observations, we have recently developed a coarse-grained molecular model of unentangled associating star-shaped polymers [8]. Given that we are trying to reproduce experimental data that goes far beyond the terminal relaxation time and the size of the molecules, it is important to keep the molecular picture as simple as possible. In our model (see Fig. 1), each star-shaped molecule is represented by its junction point, which contains all the molecular mass and friction. At the end of each arm there is a sticker that can bind to any other free sticker of the same or other molecule by means of second order reactions with rate constant k_A. The detachment of bonds between stickers occurs by means of first order reactions with rate constant $k_D = k_A/K_{eq}$, where K_{eq} is the equilibrium constant of association. Thus, all bonds are temporary or reversible. Whenever two stickers forms an intermolecular bond, an anchoring point is chosen randomly in the background for each one of the stickers involved by picking a Gaussian vector centered in the junction point of the molecule and with variance Nb^2, where N is the number of Kuhn segments in the arm and b is the Kuhn length. These junction points remain fixed in space or deform affinely with the flow (no junction fluctuations are considered in the model).

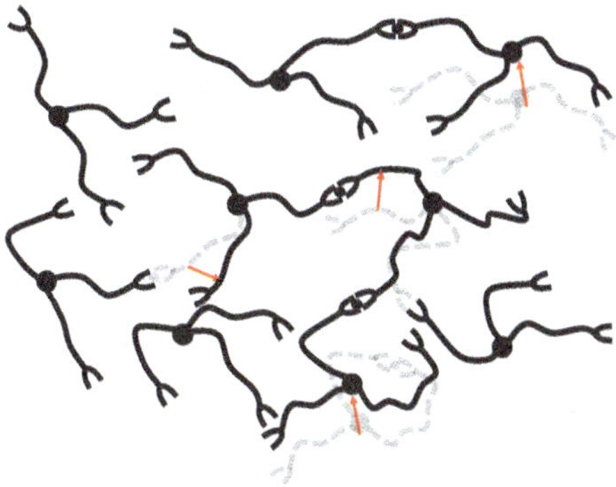

Fig. 1. Example of the molecular model used in our simulations for the case of 3-arm star-shaped molecules. In the course of a simulation, stickers form and destroy inter- and intramolecular bonds and the centers of mass move accordingly. Any individual molecule can be completely or partially attached to the percolating network, or fully detached from it.

In the simulations, we consider n star-shaped molecules with A arms each in a volume V. We track the evolution of the molecular positions by means of Brownian dynamics simulations. The units of length and time are the size $\sqrt{Nb^2}$ and the relaxation time τ_R of the arms, respectively. In these units, the stochastic differential equation of motion for the junction point of molecule i is:

$$dr_i = \frac{dt}{A}\sum_{j=1}^{A}(a_{ij} - r_i)\, l_{ij} + \sqrt{\frac{2}{A}} dW_i, \tag{1}$$

where a_{ij} is the position of the anchoring point for arm j of molecule i, W_i is a standard Wiener process and the coefficient l_{ij} is 1 if the arm i is attached to the background and 0 otherwise. We keep track of the state of all the stickers in the system by means of a stochastic chemical reaction algorithm [9], considering three reaction channels in the system: intermolecular association, intramolecular association and detachment. The number of reactions per time step Δt along each reaction channel are random numbers drawn from Poisson distributions with means given by:

$$\begin{aligned} n_{\text{INTER}} &= \frac{k_A \Delta t}{V} \sum_{i=1}^{n} f_i (F - f_i) \\ n_{\text{INTRA}} &= \frac{k_A \Delta t}{V_{\text{span}}} \sum_{i=1}^{n} f_i (f_i - 1) \\ n_{\text{DETACH}} &= \frac{k_D \Delta t}{2} (nA - F), \end{aligned} \tag{2}$$

where f_i is the number of free stickers of molecule i, $F = \sum f_i$ is the total number of free stickers in the system and $V_{\text{span}} = 4\pi(Nb^2)^{3/2}/3$ is the volume spanned by one molecule. The balance between the fraction of intermolecular and intramolecular bonds at equilibrium is controlled by the value of K_{eq} and the concentration, given by $\beta = nV_{\text{span}}/V$ (the ratio of the volume of the system and the volume spanned by the molecules). At low concentrations ($\beta \gg 1$) a high fraction of the stickers are forming intramolecular bonds (or loops) whereas

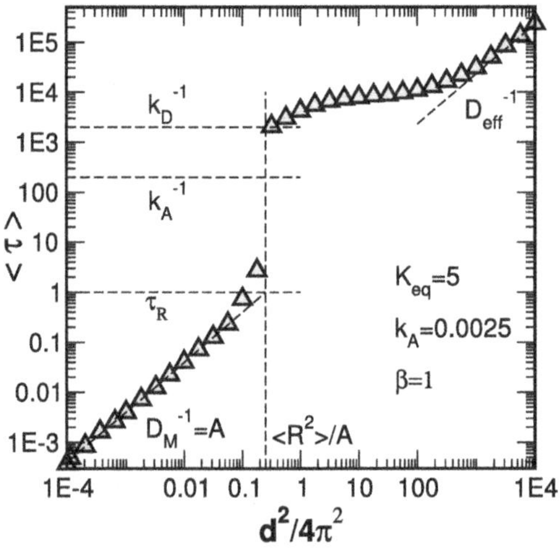

Fig. 2. Characteristic plot from a simulation of the FRS experiment on a system of 4-arm star-shaped associating polymers, showing the three diffusion regimes discussed in the main text: (i) fast Fickian diffusion at small distances, (ii) escape from the cage and superdiffusive regime at intermediate distances and (iii) slow Fickian diffusion at very long distances.

at high concentration ($\beta \ll 1$) most of the stickers are forming intermolecular bonds. For the sake of the stress tensor, only intermolecular bonds are elastically active. Here, we do not explore any other topological defects of higher order than intramolecular loops.

3 Results and Conclusions

When simulations are run at equilibrium, we can measure the one-dimensional dynamic structure factor and extract the longest average relaxation time $\langle \tau \rangle$ as a function of the scattering distance d, in a similar way as the results from an FRS experiment are obtained. On Fig. 2, $\langle \tau \rangle$ is plotted as a function of the square of d from simulations, revealing three different mechanisms of motion. At very small distances ($d^2/4\pi^2 < 1$, caging mechanism), the molecules can diffuse freely with a diffusion coefficient $D_M \propto (AN)^{-1}$ up to a distance which is of the order of the molecular size divided by A. After that they need to detach from the network before they can move any further and, for that to happen, they need to wait a time of order k_D^{-1}. Then, they can move to longer distances by a walking mechanism (detaching an arm and attaching it to a distant position) or, in the event that the molecule can detach all of its arms, it can diffuse freely over distances much longer than its radius of gyration, provided that $k_A \ll \tau_R$ (hopping mechanism). At very long distances ($d^2/4\pi^2 > 1000$), the molecules diffuse with an effective diffusion coefficient D_{eff} which is the result of walking and hopping. In the crossover from the escape of the cage to the final Fickian regime, the plot shows a phenomenological superdiffusive scaling which successfully explains the experimentally observed behavior [8]. The three regimes and the overall shape of the curve depend very strongly on polymer concentration, length of the strands between stickers and the association/dissociation kinetics. The observed superdiffusive scaling results primarily from molecular hopping, which dominates over walking when the kinetics of attachment are much slower than the relaxation time of the strands between stickers.

Acknowledgements. MB, AT and JR acknowledge funding from grant PEJ-2017-AI/IND-6767 of the Government of Madrid and project UPM-RP-160543006 from Univ. Politécnica de Madrid.

References

1. Aida, T., Meijer, E.W., Stupp, S.I.: Science **335**, 813–817 (2012)
2. Chen, Q., Zhu, L., Chen, H., Yan, H., Huang, L., Yang, J., Zheng, J.: Adv. Funct. Mater. **25**, 1598–1607 (2015)
3. Lee, K.Y., Mooney, D.J.: Chem. Rev. **101**, 1869–1880 (2001)
4. Li, Y., Rodrigues, J., Tomás, H.: Chem. Soc. Rev. **41**, 2193–2221 (2012)
5. Cordier, P., Tournilhac, F., Soulié-Ziakovic, C., Leibler, L.: Nature **451**, 977–980 (2008)
6. Tang, S., Wang, M., Olsen, B.D.: J. Am. Chem. Soc. **137**, 3946–3957 (2015)

7. Tang, S., Habicht, A., Li, S., Seiffert, S., Olsen, B.D.: Macromolecules **49**, 5599–5608 (2016)
8. Ramirez, J., Dursch, T.J., Olsen, B.D.: Macromolecules **51**, 2517–2525 (2018)
9. Gillespie, D.T.: Annu. Rev. Phys. Chem. **58**, 35–55 (2007)

Use of Rheology as a Tool for Interfacial Characterisation of Silkworm Pupae (*Bombyx mori*) Protein Concentrate Adsorbed at A/W Interface

C. Bascon[1], M. Felix[1], V. Pérez-Puyana[2(✉)], and J. de la Fuente[1]

[1] Department of Chemical Engineering, Escuela Politécnica Superior, University of Seville, Seville, Spain
carmenbsg_93@hotmail.com, {mfelix, jfferia}@us.es
[2] Department of Chemical Engineering, Facultad de Química, University of Seville, Seville, Spain
vperez11@us.es

Abstract. The use of insect proteins for food applications has been boost by the current preferences of consumers who prefer healthier diets. Although Silkworm (SLW) pupae (*Bombyx mori*) has been used in food products, their use is not widespread in western countries. Interfacial properties of silkworm pupae (*Bombyx mori*) protein concentrate were studied at two pH values (2.0 and 8.0). The analysis of the protein adsorbed at A/W interface was carried out using dilatational and interfacial shear measurements. Dilatational measurements were carried out using a pendant droplet tensiometer (Tracker, Teclis Scientific), whereas interfacial shear measurements relied on a rheometer (DHR-3, TA Instruments). Results obtained confirmed the suitability of SLW protein to stabilize the interface. Moreover, at pH 8.0 proteins were adsorbed faster at the A/W interface, whereas the rheological response at pH 2.0 exhibited higher viscoelastic moduli and lower frequency-dependence.

1 Introduction

Due to the rapid increase in world population, the demand for protein derived from animals will double, with a portion of this protein demand being satisfied by the current protein sources. However, the International Livestock Research Institution predicts that at least 70% of this demand must be satisfied with innovative technologies and new protein sources [1]. Insects are one of the new proteins sources that can help to satisfy this growing demand [2]. Moreover, insects not only are a source of high-nutritional value proteins, but also they have a lower environmental impact [3]. Despite increasing consumption (due in part to the legalization of insects in Europe), insects currently do not contribute significantly to the human diet in developed countries. A recent study carried out in the Netherlands, Australia and Germany suggested the introduction of "invisible insects" (with processed insect protein) into food products could be a way to improve consumer acceptance in countries where their intake is not traditional [4]. Many foods are presented in the form of dispersed systems such as emulsions and

F. J. Galindo-Rosales et al. (Eds.): IBEREO 2019, SPM, pp. 123–126, 2020.
https://doi.org/10.1007/978-3-030-27701-7_26

foams. Researchers are focused on the development of kinetically stable systems over a reasonable period of time by including surface agents. Proteins are among the main agents with interfacial properties that are used in food products. In addition to their participation in the preparation and stabilization of the final system, proteins also contribute to increase the nutritional quality of the food product, as well as to the acquisition of the desirable sensory characteristics such as texture, structure or flavor.

In the present work the functional properties of the protein contained in the silkworm pupae (*Bombyx mori*) were determined by the characterization of Interfacial properties through dilatational tests and interfacial shear rheology for its use as a food stabilizer. Both tests were carried out at pH 2.0 and 8.0.

2 Materials and Methods

2.1 Materials

The silkworm protein concentrate (SLW) used in this study was supplied by FeedStimulants (the Netherlands), containing up to 50.5 ± 0.3 wt% proteins. Chemical reagents (i.e. HCl, NaOH, NaH_2PO_4) were purchased from Sigma–Aldrich company (St. Louis, USA). The solutions were prepared using Milli-Q grade water. SLW solutions were prepared at 1 wt% soluble protein (0.1 mg/mL) in 50 mM phosphate and Tris-Base buffers (pH 2.0 and 8.0).

2.2 Dilatational Measurements

Dilatational measurements were carried out in a droplet tensiometer (Tracker, Teclis scientific, France). Oscillatory dilatational experiments were carried out during the protein adsorption (at 0.1 Hz) and after 10,800 s (from 0.075 Hz to 0.1 Hz). All experiments were carried out in triplicate at 20.0 ± 0.1 °C and using a low absorbance glass cuvette (8 ml).

2.3 Interfacial Shear Measurements

Interfacial shear measurements were carried out using a DHR-3 rheometer (TA Instruments, New Castle, USA), with a double-wall-ring geometry (DWR). Interfacial viscoelastic moduli were obtained during the protein adsorption (at 0.1 Hz) and after reaching the quasi-equilibrium state (10,800 s) from 0.0075 to 0.3 Hz. All these experiments were carried out within the linear viscoelastic region (LVR) at 20.0 ± 0.1 °C.

3 Results and Discussion

Viscoelastic properties of SLW obtained during protein adsorption at A/W interface ($t < 180$ s), with dilatational (A) and interfacial shear (B) measurements, are shown in Fig. 1. Regardless of the technique used, the adsorption of SLW protein at pH 2.0 takes longer than at pH = 8, showing higher complex dilatational moduli (E^*) when the interface is characterized by dilatational measurements. This result could be related with the fact that interfacial shear measurements are more sensitive to protein-protein interactions [5].

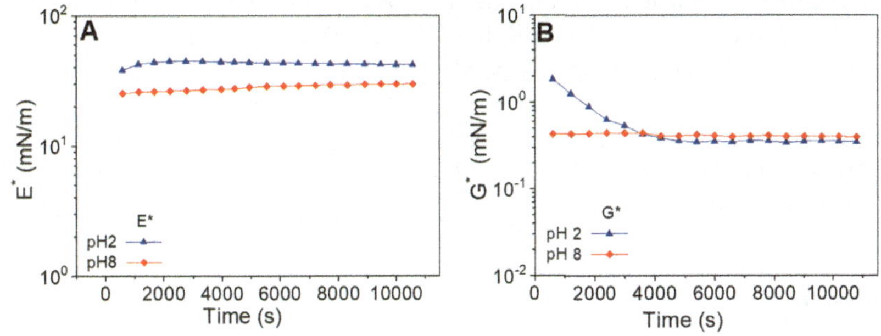

Fig. 1. Viscoelastic properties of SLW during protein adsorption at A/W interface (t < 180 s), obtained from dilatational (A) and interfacial shear (B) measurements.

Both, dilatational and interfacial shear techniques offer complementary information. Whereas the former shows an increase in the complex moduli, the later exhibits a decrease at pH 2.0 and a nearly constant value at pH 8.0, suggesting that structural changes take place at pH 2.0 during protein adsorption at A/W interface. Once the SLW protein system was adsorbed and the pseudo-equilibrium state was reached, the mechanical spectra of the interface were obtained. Figure 2 shows E* and G* as a function of frequency from dilatational (A) and interfacial shear (B) measurements.

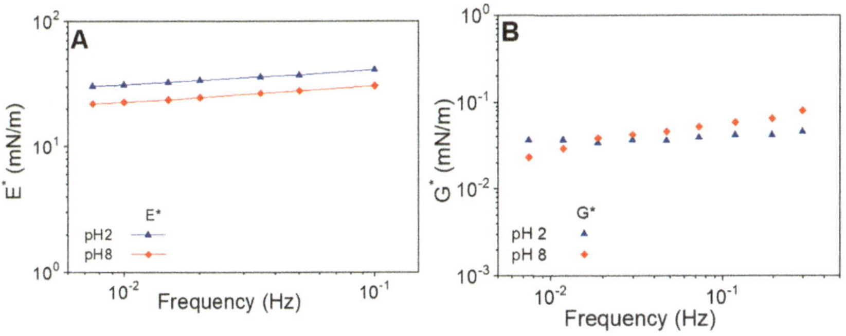

Fig. 2. Mechanical spectra of interfaces stabilized after SLW protein adsorption at A/W interface (t > 180 s), obtained from dilatational (A) and interfacial shear (B) measurements.

Results obtained from the mechanical spectra showed a lower frequency-dependence when the interface is characterized by interfacial shear measurements. Furthermore, when the interface was characterized by dilatational measurements, higher E* values were obtained at pH 2.0.

4 Conclusions

Obtained results indicate that SLW protein systems can be regarded as an alternative protein source which could be used for the stabilization of dispersed systems such as foams. The results indicated that the interfacial layer formed at pH 2.0 is stronger than the interfacial film developed at pH 8.0. These results confirm that the functional properties of the proteins strongly depend on the pH evaluated.

Acknowledgements. Authors also acknowledge for the pre-doctoral fellowship of Víctor M. Pérez Puyana (VPPI-US) and the post-doctoral fellowship of Manuel Félix Ángel (VPPI-US) to University of Seville.

References

1. Armstrong, W.D.: Livestock Research Institution (2009)
2. Rumpold, B.A., Schlüter, O.K.: Mol. Nutr. Food Res. **57**(5), 802–823 (2013)
3. Oonincx, D.G.A.B., de Boer, I.J.M.: PLoS ONE **7**(12), e51145 (2012)
4. Hartmann, C., Siegrist, M.: ACR North American Advances, pp. 41–44 (2016)
5. Leiske, D.L., Raju, S., Ketelson, R., et al.: Exp. Eye Res. **90**(5), 598–604 (2010)

Photo Rheometry of Waterborne Polyurethanes for Its Implementation in Vat Photopolymerization

Robert Hernández Aguirresarobe(✉), Fermín Elizalde Iraizoz, Haritz Sardon, and Antxón Santamaría

POLYMAT and Polymer Science and Technology Department, Faculty of Chemistry, University of the Basque Country (UPV/EHU), Avda. Tolosa 72, 20018 San Sebastian, Spain
{roberto.hernandez, haritz.sardon, antxon.santamaria}@ehu.eus, fermin.elizalde@polymat.eu

Abstract. The introduction of additive manufacturing (AM) techniques is revolutionizing the field of polymer processing, as it provides a unique alternative to produce three-dimensional shapes and highly personalized structures. [1] Among different (AM) techniques, Vat photopolymerization, material jetting and binder jetting technologies rely on photo cross-linkable resins. Thus, Photo rheometry has gained attention as it appears as an essential tool to understand the material behaviour in the process and the development of new feedstock materials. In essence, photo rheometry measurements permit to analyse rheological behaviour of materials during photochemical processes, mainly cross-linking reactions.

In this work, we present the analysis of the rheological performance of waterborne polyurethanes containing acrylate groups as a proof of concept for its incorporation as feedstock materials for Vat photopolymerization [2].

1 Introduction

The use of light as energy source for producing chemical reactions has been an issue in polymer chemistry for decades. In fact, most of polymerization reactions in industry are carried out by incorporating photoinitiators to chemical reaction mixtures, where the most common use of photoinduced reactions are the ones for acrylic based monomers, such as acrylates styrene, olefins and so on. Thus, the control of the rheological properties of reaction mixtures during the irradiation processes is essential in industrial applications, especially in those systems susceptible to produce material cross-linking. This necessity gave rise to the development of photo rheometry devices capable of following the evolution of the viscoelastic properties during material irradiation.

Although this characterization is not new, photo rheometry has gained attention since the irruption of Vat photopolymerization additive manufacturing technologies. [3] This technology is capable of forming well-defined three-dimensional shapes by crosslinking photo resins layer-by-layer [2]. The printing speeds and physical

F. J. Galindo-Rosales et al. (Eds.): IBEREO 2019, SPM, pp. 127–131, 2020.
https://doi.org/10.1007/978-3-030-27701-7_27

performance of the final specimens (defined by the interlayer welding during the curing) is directly related with the rheological properties of materials upon irradiation. Thus, photo rheometry has become an essential tool in the material development for this technology.

The development of specific materials for 3D printing technologies is not trivial. In fact, the material portfolio for additive manufacturing technologies is very scarce and far from traditional polymer commodities. In the specific case of Vat photopolymerization, this kind of materials are mainly based on acrylic monomer and oligomer mixtures in water. In this work, we propose the incorporation of waterborne polyurethane structures as feedstock materials for Vat photopolymerization. The photo rheological performance of synthesized polyurethanes, containing acrylic groups in the main backbone of the polymer will be analysed using photo rheometry and its suitability will be compared with commercially available resins.

2 Materials and Methods

Polyurethanes, named here as HEMA5, were synthesized following the procedure described elsewhere [3] and summarized in Fig. 1.

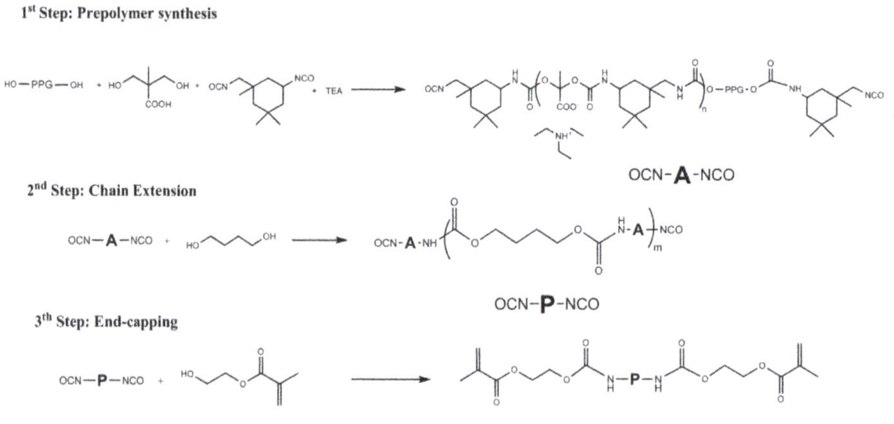

Fig. 1. Synthetic pathway used for the waterborne polyurethane production.

HEMA 5 polyurethanes are synthesized using isophorone diisocyanate, polypropylene glycol, 2,2-dimethylolpropionic acid and butanediol as monomers. Polyurethanes were end-capped with hydroethyl methacrylate in order to have a radically cross-linkable material. These materials were further polymerized by introducing photoiniciators and applying UV light. Photopolymerization were carried out using 1 wt% of 2,2-dimethoxy-phenylacetophenone as photoinitiator. A commercial photopolymer resin B9R-1-RED (B9Creations) was also studied for comparison purposes. Photo rheometry measurements were carried out in an AR-G2 rheometer equipped with a UV irradiation accessory (365 nm), in parallel plate geometry, at room temperature and irradiation conditions described in the text. Additionally, sample viscosity analysis was carried out in an Anton Paar MC 101 rheometer using concentric cylinder geometry.

3 Results and Discussion

We first studied the intrinsic capabilities of polyurethanes to form three-dimensional networks upon UV irradiation in the presence of a photoinitiator. This study was carried out by performing time sweep experiments under 2 different irradiation intensities: 10 and 100 mW/cm^2. The obtained results are depicted in Fig. 2.

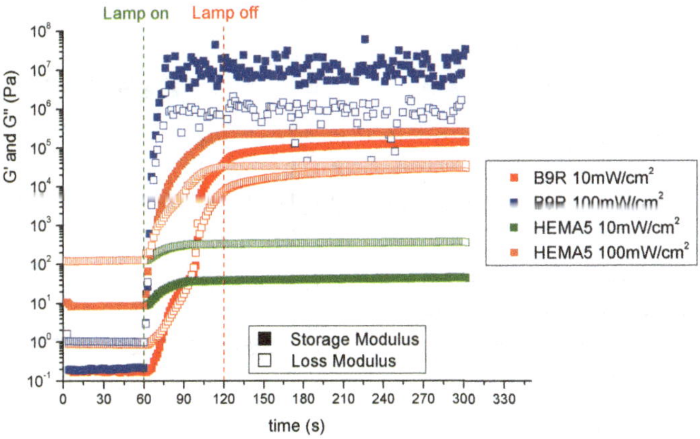

Fig. 2. Cross-linking behaviour of HEMA based systems and reference B9R resin at different irradiation intensities. Experiments were performed at 25 °C.

Non-irradiated materials showed a predominantly viscous behaviour, characterized by a loss modulus larger than the elastic modulus ($G'' > G'$). Once the lamp was switched on the both G' and G'' moduli started to increase and a cross-over could be observed in the case of 100 mW/cm^2 irradiated sample. From that point on, $G' > G''$ indicating the cross-linked nature of the formed material. However, 10 mW/cm^2 light intensity was not able to produce the material cross-linking as no cross-over was observed. Even so, the cross-linking kinetics shown by the proposed polyurethanes are much lower than the one for the commercial resin, which is defined by a sharp increase in both moduli within a few seconds and cross-linking taking place at both light intensities.

Another key parameter when developing materials for Vat photopolymerization is the viscosity of materials, as it determines the capacity of the resin to renew the feedstock material once the platform moves to form a new layer. Thus, to ensure this renewing process, resins should be low viscosity polymer solutions and/or dispersions. The viscosity of non irradiated materials was measured for dried, acetone solution (30 wt%) and water dispersion (30 wt%) (Fig. 3a). Additionally, the viscosity of the commercial resin was included for comparison purposes. As shown, dried polyurethane showed high viscosity values, which can potentially impede its implementation as photopolymer resin. However, both the acetone solution and water dispersion show lower viscosity values in the range of commercial resin (even lower), with the acetone solution the one presenting the higher viscosity values. However, the presence of solvent strongly affects the viscoelastic characteristic of the material upon irradiation (Fig. 3b). Thus, neither the acetone solution nor the waterborne polyurethane dispersion showed G′ G″ cross-over, revealing the unsuccessful capacity to form a three-dimensional network. This contrasting behaviour of polyurethane dispersion in relation with commercial resins can be attributed to the higher monomer concentration presented in such resins.

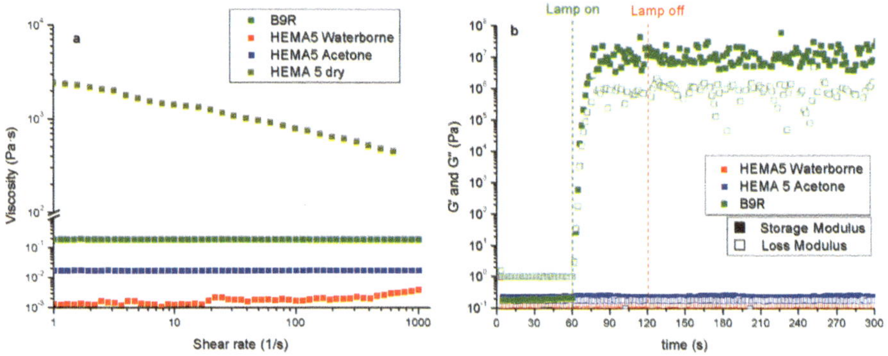

Fig. 3. Viscosity vs. shear rate curves for different materials (a) and cross-linking behaviour upon irradiation for different dispersed systems (b).

4 Conclusions and On-going Work

The study shows the importance of the photo rheometry in the development of resins for Vat photopolymerization. The results reveals the potential capacity of acrylic end-capped polyurethanes to be used as photo crosslinkable resins for Vat photopolymerization, although several drawbacks have to be faced. First, the cross-linking capacities of polyurethanes show slower kinetics than commercial resins, therefore there are not successful in forming a polymer network. In addition, the low HEMA concentration used in the sample is insufficient to produce three-dimensional networks in the form of water dispersions.

As possible alternatives for the on-going work, we suggest the incorporation of water soluble photoinitiators, such as quaternary salts (Irgacure 2959) and increasing the HEMA amount in the samples, thus reducing the molecular weight but increasing the functionality of the final resin.

Acknowledgements. This research was funded by MINECO (MAT2017-84116-R).

References

1. Wilts, E.M., Pekkanen, A.M., White, B.T., Meenakshisundaram, V., Aduba, D.C., Williams, C.B., Long, T.E.: Vat photopolymerization of charged monomers: 3D printing with supramolecular interactions. Polym. Chem. **10**, 1442–1451 (2019)
2. Tofail, S.A.M., Koumoulos, E.P., Bandyopadhyay, A., Bose, S., O'Donoghue, L., Charitidis, C.: Additive manufacturing: scientific and technological challenges, market uptake and opportunities. Mater. Today **21**, 22–37 (2018)
3. Llorente, O., Fernández-Berridi, M.J., González, A.I.: Study of the crosslinking process of waterborne UV curable polyurethane acrylates. Prog. Org. Coat. **99**, 437–442 (2016)

Rheometry and Experimental Methods

Assessing the Sliding Cylinder Approach to Determine Instantaneous Viscosity Under Unsteady Flow Conditions

Ahmad Fakhari[1](✉) and Francisco J. Galindo-Rosales[2]

[1] CEFT, Departamento de Engenharia Mecânica, Faculdade de Engenharia da Universidade do Porto, Rua Dr. Roberto Frias, s/n., 4200-465 Porto, Portugal
ahmadfakhari@gmail.com

[2] CEFT, Departamento de Engenharia Química, Faculdade de Engenharia da Universidade do Porto, Rua Dr. Roberto Frias, s/n., 4200-465 Porto, Portugal
galindo@fe.up.pt

Abstract. This preliminary numerical investigation is carried out in order to assess the usefulness of a new variant of the sliding cylinder approach to determine "instantaneous" viscosity under unsteady flow conditions. A fixed cylindrical reservoir is considered with a coaxial inner cylinder which is sliding into the liquid with different time dependent velocity profiles, i,e. a constant velocity, linear velocity and parabolic profile. For each velocity profile, three different blockage ratios (the ratio of the impactor's diameter over the reservoir's diameter) of 1/1.5, 1/3 and 1/6 are considered.

OpenFOAM 2.4.0 is used to simulate the fluid flows, and the multiphaseInterDyMFoam solver is employed to handle dynamic mesh for solving the Navier-Stokes equations of two phases flow consisting of air and the liquid. The simulations start with the sliding cylinder's tip at the interface of the two phases (liquid-air), and it moves toward the liquid with the prescribed velocity profile. In order to assess the usefulness of this approach to perform rheological measurements at short acquisition times, we focus our attention on the fluid-structure interaction between the sliding cylinder and the liquid, mainly the contributions of drag and pressure force to the total force exerted by the fluid to the piston.

1 Introduction

Modern rotational rheometers are extremely sensitive devices able to provide accurate measurements of different material functions of a wide variety of complex materials. Probably due to scientific and industrial interest, these rheometers have been intensively improved in such a way that they are able to impose/measure low deformations/torques. Nevertheless, they present serious limitations when they try to characterise complex fluids under "large, rapid,

F. J. Galindo-Rosales et al. (Eds.): IBEREO 2019, SPM, pp. 135–138, 2020.
https://doi.org/10.1007/978-3-030-27701-7_28

transient shear deformation" and with data acquisition times of the order of milliseconds, either because of instrument or fluid inertia issues [1].

Dealy and Giacomin [2] discussed about the pros and cons of using a sliding cylinder rheometer, being the main advantage the rectilinear flow that helps in overcoming some of the limitations of rotational rheometers. Moreover, they also listed the main applications in which sliding cylinder rheometers were used, having a force transducer in the shaft of the inner cylinder; nevertheless, they did not mention any application for ultrashort timescales and to the best of the authors knowledge, the sliding cylinder rheometer was never tested for measuring the viscosity at short acquisition times, below 50 ms.

This preliminary study is purely numerical and it aims at providing some insight about the experimental conditions under which the sliding cylinder rheometer could provide reliable results to obtain the "instantaneous" viscosity curve of a fluid in just 40 ms.

2 Materials and Methods

2.1 Geometry and Boundary Conditions

The two cylinders are coaxial, being the outer one fixed ($BR = d/D = 1/1.5$, $1/3$ and $1/6$ respectively) and the inner one ($d = 16$ mm) moves with different velocity profiles (Fig. 1a): $v(t) = 1.2$ m/s (constant); $v(t) = 1.2 - 29.09t$ m/s (linear); and $v(t) = 705t^2 - 1.84 \cdot 10^{-4}t + 1.2$ m/s (parabolic). The tip of the impactor is initially submerged in the liquid a distance z_0, being $z_0 = 0,\ d,\ 2d,\ d^2$. No-slip boundary condition with zero velocity at the walls was imposed, except at the impactor given by $v(t)$. Zero pressure gradient boundary condition was applied on all walls.

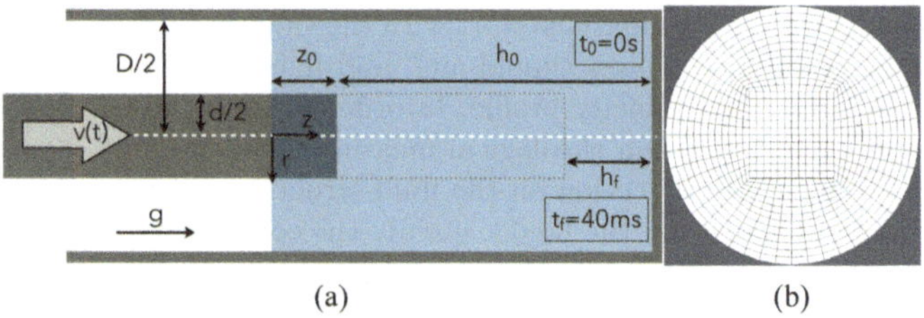

Fig. 1. (a) Sketch of the sliding cylinder rheometer, $z_0 = d$. (b) Cylinder's cross-section mesh.

2.2 Numerical Method

OpenFOAM 2.4.0 was used to simulate the time dependent 3D fluid flows. The multiphaseInterDyMFoam solver was employed to handle dynamic mesh for solving Navier Stokes equations of two phases flow consisting of air ($\eta = 1.8375 \cdot 10^{-6}$

Pa·s and $\rho = 1.225$ kg/m^3) and the liquid ($\eta = 10^3$ Pa·s and $\rho = 970$ kg/m^3). Cylindrical polar coordinates were considered (r, z, θ). The structured mesh was generated using blockMesh utility. While the grid is uniform in the radial direction, a constant stretch ratio was applied in the stream-wise direction (z) to have highest resolution at the impactor's tip, where there is air/liquid interface for $z_0 = 0$. The simulations were carried out using a fixed time step $\Delta t = 1.375$ μs, with the simulation time equal to 41 ms. Euler time integration was applied and the data has been printed every 1.375 ms. Gauss linear discretization scheme was used for the derivatives discretization. The drag force $(F_w = \tau_w \cdot A_L)$, due to the friction exerted by the liquid at the lateral wall of the inner cylinder

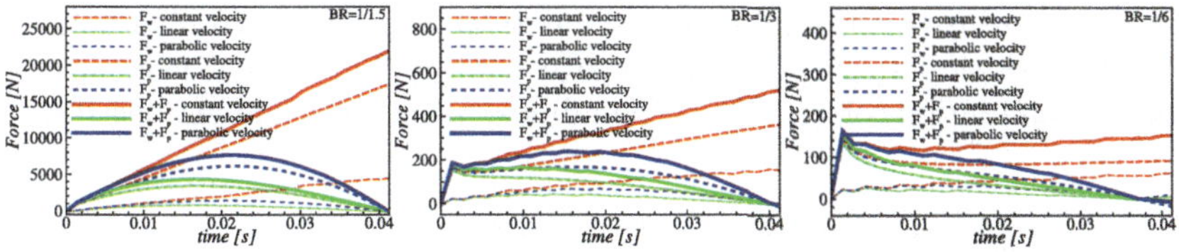

(a) BR=1/1.5 (b) BR=1/3 (c) BR=1/6

Fig. 2. Drag force versus pressure force ($z_0 = 0$).

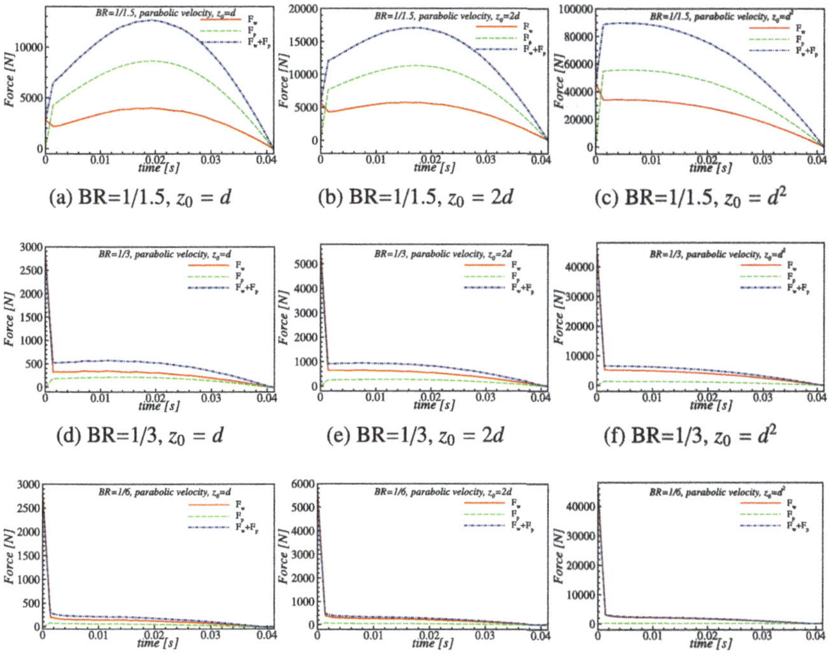

(a) BR=1/1.5, $z_0 = d$ (b) BR=1/1.5, $z_0 = 2d$ (c) BR=1/1.5, $z_0 = d^2$

(d) BR=1/3, $z_0 = d$ (e) BR=1/3, $z_0 = 2d$ (f) BR=1/3, $z_0 = d^2$

(g) BR=1/6, $z_0 = d$ (h) BR=1/6, $z_0 = 2d$ (i) BR=1/6, $z_0 = d^2$

Fig. 3. Drag force versus the pressure force for impactor with parabolic velocity at different blockage ratio and different $z_0 \neq 0$.

($\tau_w = \eta \cdot \dot{\gamma}_w$), is computed at every time step and compared with the pressure force at the bottom of the inner cylinder ($F_P = P \cdot \pi \frac{d^2}{4}$).

3 Results and Discussion

For rheometric purposes, it is preferable to impose a parabolic velocity profile than constant or linear, because F_w is larger (Fig. 2). Moreover, lower BR improves the contribution of the drag force. In Fig. 3 it can be observed that the larger the value of z_0, the more dominant will be the contribution of the drag force with regards to the pressure force, although a higher added-mass effect is present at the early stages of the simulations.

A more extensive set of numerical simulations of other Newtonian fluids with different viscosities are currently under investigation. Later, we expect to develop numerical simulations of non-Newtonian fluids, using GNF and viscoelastic models.

References

1. Ewoldt, R.H., Johnston, M.T., Caretta, L.M.: Experimental challenges of shear rheology: how to avoid bad data. In: Spagnolie, S. (eds.) Complex Fluids in Biological Systems. Biological and Medical Physics, Biomedical Engineering. Springer, New York (2015)
2. Dealy, J.M., Giacomin, A.J.: Sliding plate and sliding cylinder rheometers. In: Collyer, A.A., Clegg, D.W. (eds.) Rheological Measurement. Springer, Dordrecht (1998)

Large Amplitude Oscillatory Shear (LAOS) Experiments on Colloidal Ceramic Paste Formulated for Robocasting Applications

Bo Nan[1], Francisco J. Galindo-Rosales[2], and José M. Ferreira[3](✉)

[1] CEITEC - Central European Institute of Technology, Brno University of Technology, 612 00 Brno, Czechia
Bo.Nan@ceitec.vutbr.cz

[2] CEFT, Department of Chemical Engineering, Faculty of Engineering of the University of Porto, 4350-465 Porto, Portugal
galindo@fe.up.pt

[3] Department of Materials and Ceramic Engineering, University of Aveiro, CICECO – Aveiro Materials Institute, 3810-193 Aveiro, Portugal
jmf@ua.pt

Abstract. As a branch of direct 3D printing technique, Robocasting is a promising candidate to fulfil multiple material printing [1]. Currently, the Robocasting community commonly utilizes small amplitude oscillatory shear (SAOS) tests to evaluate the printability and stiffness of the pastes for Robocasting, while large amplitude oscillatory shear (LAOS) test has been neglected. However, SAOS experiments restrict the information about the rheological properties within the linear viscoelastic regime, providing information regarding the internal microstructure of the pastes at rest, i.e., before they start to flow. From the previous study on large amplitude oscillatory shear (LAOS) tests of polymer solutions and powder suspensions [2, 3], the Lissajous–Bowditch curves illustrate the gradually changing flow behaviours of those complex fluids. By imposing variable frequencies and shear strains to inks with both high elastic and viscous moduli, the same approach is used to assess their complex theological behaviours, aiming at further understanding their effects on the Robocasting process.

1 Introduction

Robocasting is an ink-based 3D printing technique widely used in material engineering [1] and has recently become a hot topic in tissue engineering [4]. The printability of the ink is an important factor for the printing process and is usually related to its rheological properties. In our previous report [5], a series of aqueous based inks were prepared by combining suitable proportions of lead-free piezoelectric ceramic powder, dispersant, binder and coagulating agent. The viscoelastic properties of the inks measured under oscillatory tests revealed that the most influencing factors were the quantity of the ceramic powder (solid-loading) and the concentrations of the coagulating agent (polyethyleneimine, PEI). Therefore, it may occur that, taking titanium oxide as an example, two pastes with totally different compositions (T17 and T18)

F. J. Galindo-Rosales et al. (Eds.): IBEREO 2019, SPM, pp. 139–143, 2020.
https://doi.org/10.1007/978-3-030-27701-7_29

shown in Table 1, have both elastic modulus (G') and loss modulus (G'') of the same magnitude in the linear viscoelastic region (shown in Fig. 1). These two almost overlapped curves may obscure the real behaviour of the ink and mislead the printing process, if the tests were performed only in the linear viscoelastic region. To reveal the differences between the two inks with similar results within the linear viscoelastic region, the ceramic inks containing titanium oxide are also measured under LAOS for further investigation.

Table 1. Chemical compositions of the titanium oxide inks

Number	Final solid-loading (vol.%)	Dispersant (wt%)	HPMC (wt%)	PEI (wt%)
T16	38	0.8	2.0	0.030
T17	38	0.8	2.0	0.045
T18	42	0.8	2.0	0.030

2 Materials and Methods

2.1 Preparation of the Samples

Aqueous stock solutions of hydroxypropyl methylcellulose (HPMC, Sigma-Aldrich, USA, M_n ~10,000 – binder) and of polyethyleneimine (PEI, Sigma-Aldrich, USA, M_n ~1,800 and M_w ~2,000 – coagulating agent) were prepared in advance. The polymer solution concentrations were 33.0 wt% for HPMC and 5.0 wt% for PEI. The formulations of both inks (T17 and T18) are shown in Table 1, together with T16 ink, used as the control group. Suspensions with final solid-loadings of 38 and 42 vol.% were prepared by mixing deionized water with 0.8 wt% Dolapix CE64 (ZSCHIMMER & SCHWARZ, Germany) and titanium oxide powder (Riedel-de Haën, Germany, D_{50} = 0.207 μm). Then, 2.0 wt% of binder was added to increase the intrinsic viscosity of the dispersing liquid. Accordingly, inks with sufficient stiffness could be obtained by further adding 0.03 or 0.045 wt% of PEI. The suspensions were homogenized in the planetary mixer for 5 min at the rate of 600 rpm (with only dispersant) and at 1300 rpm (after successively adding the binder and the coagulating agent). The percentages of processing additives were calculated based on the initial mass of solids.

2.2 Rheological Analyses

The viscoelastic properties of the inks were characterized by a stress-controlled rotational rheometer (Anton Paar MCR301, Austria) by using a direct strain oscillation module (DSO) and applying amplitude strain sweeps at the following constant frequencies of 0.6, 0.9, 1.0, 1.2 and 2.4 Hz, respectively. Measurements were performed with a serrated plate-and-plate geometry (PP10/P2, Anton Paar, Austria) with a gap of 0.4 mm. The temperature was set at 25 °C and a solvent trap cover was used to maintain the temperature and prevent water evaporation from the samples during the tests.

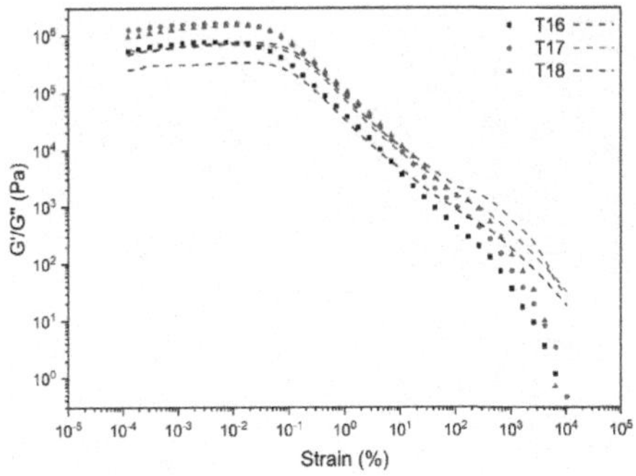

Fig. 1. Elastic moduli (G'-solid dots) and loss moduli (G''-dash lines) of the inks measured under shear strain ranging from 10^{-4} to 104% at 1 Hz.

3 Results and Discussion

Figure 1 shows that within the linear viscoelastic region (γ from 10^{-3} to 10^{-1}%), the G' curves of T17 and T18 overlap, while the G' curve of T16 is below these two curves in the entire shear strain range due to lower contents of coagulating agent and solids. After the crossover points ($\gamma \sim 10\%$), the G'' curve of T18 is higher than the other two curves before γ reaches 4 000%, which suggests that it may dissipate more energy than the other two inks under such a high shear strain.

For further understanding the influence of an increasing γ on the inks' rheological behaviour, the experimentally measured LAOS responses are shown from Fig. 2(a) to (f). The shapes of the Lissajous curves agree with the trend reported in [6] that the inks are less dissipative at low shear strains, while at higher shear strains, the inks become more dissipative. However, for $\gamma = 100\%$, the T18 dissipates less than the other two inks as shown in Fig. 2(d), which seems contradictory considering the higher G'' of the T18 ink in comparison to those of the other two inks as indicated in Fig. 1. It proves that the analysis based on the first harmonic contribution (G' or G'') are not applicable within the non-linear viscoelastic region. Instead, the Lissajous curves can provide more valuable information.

In Robocasting, the shear strain varies, depending on the nozzles and syringes utilized in the experiment, but this obviously happens within the non-linear viscoelastic region involving large and rapid deformations. Therefore, when the shear strain is as high as 10 000%, the shapes of the curves change to distorted rectangles. Even though the results obtained at higher shear strains are not reliable due to the high noise to signal ratio, which could be a limitation for further simulating the printing process, the Lissajous curves positioned in a Pipkin space can still help to understand the effects of both the frequency and shear strain applied on the ink. As shown in Fig. 3, the Lissajous curves for the T17 ink system gradually flatten. Under the same shear strain, the shape of the curves is similar, while the change in the area of the enclosed shapes represents the impact from the frequency, showing an elastic behaviour at high frequency and low shear strain, and a dissipative behaviour at low frequency and high shear strain.

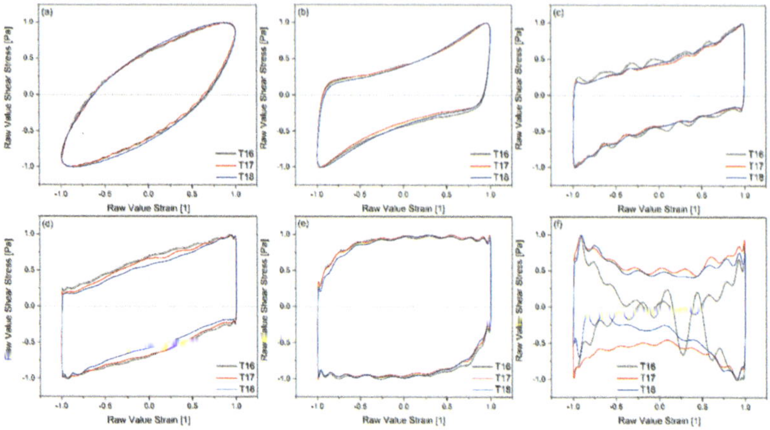

Fig. 2. The Lissajous curves [stress (y-axis) vs. strain (x-axis)] of the inks (T16, T17 and T18), arranged from small strain amplitude to large strain amplitude (a) 0.1% (b) 1% (c) 10% (d) 100% (e) 1 000% (f) 10 000% at a fixed frequency of 1 Hz.

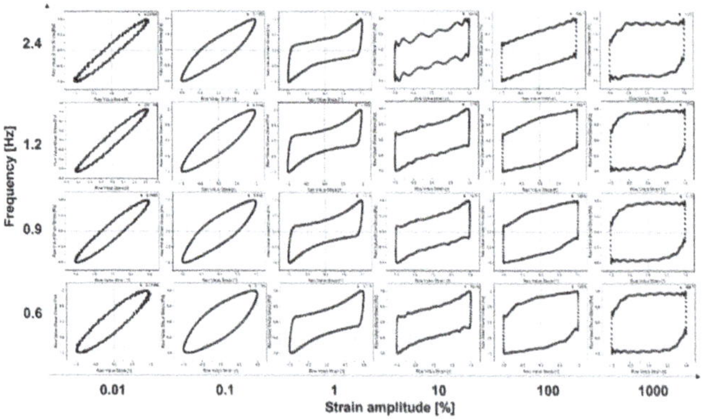

Fig. 3. Raw LAOS data for ink T17, shown as normalized Lissajous curves of shear stress $\sigma(t)$ *vs.* strain $\gamma(t)$ with an increasing shear strain and frequency.

4 Conclusions

The information from LAOS experiments can distinguish the two inks with similar SAOS response and provide more evidences on how the colloidal inks behave within the non-linear viscoelastic region. LAOS can be used to predict the combined effects of varying the shear strain and frequency on an ink during printing and also to facilitate the ink-loading into a syringe without bubbles' entrapment, which can be a guide for the future work.

References

1. Truby, R.L., Lewis, J.A.: Printing soft matter in three dimensions. Nature **540**, 371–378 (2016)
2. Ewoldt, R.H., Hosoi, A.E., McKinley, G.H.: New measures for characterizing nonlinear viscoelasticity in large amplitude oscillatory shear. J. Rheol. **52**, 1427–1458 (2008)
3. Palcevskis, E., Jakobsons, E., Faitel'son, L.: Rheological properties of nanosized AlN powder suspensions for advanced ceramics. Mech. Compos. Mater. **36**, 501–508 (2000)
4. Lewis, J.A., Kolesky, D.B., Skylar-Scott, M.A., et al.: Method of printing a tissue construct with embedded vasculature. U.S. Patent. Application No. 16/143,050
5. Nan, B., Olhero, S., Pinho, R., et al.: Direct ink writing of macroporous lead-free piezoelectric $Ba_{0.85}Ca_{0.15}Zr_{0.1}Ti_{0.9}O_3$. J. Am. Ceram. Soc. **102**, 3191–3203 (2019)
6. Hyun, K., Wilhelm, M., Klein, C.O., et al.: A review of nonlinear oscillatory shear tests: analysis and application of large amplitude oscillatory shear (LAOS). Prog. Polym. Sci. **36**, 1697–1753 (2011)

Searching for Rheological Conditions for FFF 3D Printing with Flexible Polymers

I. Calafel[1(✉)], R. H. Aguirresarobe[1], M. I. Peñas[1], A. Santamaría[1], M. Boix[2], J. I. Conde[2], and B. Pascual[2]

[1] POLYMAT and Polymer Science and Technology Department, Faculty of Chemistry, UPV/EHU, Avda. Tolosa 72, 20018 San Sebastian, Spain
itxaso.calafel@ehu.es

[2] Innovation and Technology Department, Chlorine Derivatives Division, ERCROS S.A., Diagonal 595, 08014 Barcelona, Spain

Abstract. Compounds based on PVC, typically employed to elaborate profiles, films, sheets and pond membranes by calendering, extrusion or injection processing methods, have been reformulated to obtain flexible materials apt to Fused Filament Fabrication (FFF) 3D Printing. Suitable filaments were obtained in Haake MiniLab extruder (Thermo Fisher Scientific) at a temperature of 140 °C. Since the occurrence of filament buckling depends on the ratio of the compression modulus, K, to the viscosity, η, the compounds were characterized under the premises of the printing temperatures and velocities. From the low K values found for the filaments it was inferred that buckling would arise during 3D Printing. However, FFF process was found to be perfectly feasible for selected samples. A rheological explanation of this unexpected and positive result was proposed considering the hypothesis of a plug flow in the nozzle.

1 Introduction

At least 60% of the materials currently used in additive manufacturing, most commonly termed as 3D printing, are polymers. These materials are unique for the very widely employed Fused Filament Fabrication (FFF) method, which consists in feeding a reservoir with a solid filament which is molten in the process and extruded through a needle in a moving platform. So far the most popular polymers utilized for FFF are polylactic acid (PLA) and acrylonitrile-butadyene-styrene (ABS) copolymer, but the inclusion of other polymers is being considered day by day. An important issue in FFF is the occurrence of buckling of the filament before the polymer passes from the solid to the molten state. This typically happens in amorphous polymers, because their compressibility modulus (K) is not high enough to give a proper pressure to the polymer melt to flow through the needle. Actually, the parameter K/η, where η is the viscosity, determines the feasibility of a polymer for FFF: The lower the value of K/η is, the higher the risk of buckling. In this framework, flexible polymers, which possess a low K value, are not good candidates for FFF and, in particular, poly(vinyl chloride) (PVC) plasticised polymers could be a priori discarded. However, in this work we demonstrate that, in fact, flexible PVCs can be used in FFF 3D printing, leading to an apparent contradiction, which is discussed in this work.

F. J. Galindo-Rosales et al. (Eds.): IBEREO 2019, SPM, pp. 144–147, 2020.
https://doi.org/10.1007/978-3-030-27701-7_30

2 Materials and Methods

The characteristics of the PVC compounds studied in this work are presented in Table 1. These are commercial grade polymer based in Etinox® 650 and Etinox® 630, both of them supplied by Ercros S.A. Those resins are used typically to elaborate profiles, films, sheets and pond membranes by calendering, extrusion or injection processing methods. In this case, both PVC resins have been plasticised by 40 phr of dioctyl terephthalate (DOTP).

Table 1. Properties of the PVC plasticized compounds.

	PVC-1	PVC-2
Type of resin	Etinox® 650	Etinox® 630
Shore hardness (Type A)	86	85
Shore hardness (Type D)	-	-
Density (g/cm^3)	1.25	1.25
Tensile strength (MPa)	28	22
Elongation (%)	387	344
Compression Modulus (MPa)	39	34

Viscosity as a function of shear rate, $\eta(\dot{\gamma})$, results were obtained by capillary rheometry in a Göttfert 2002 rheometer (Buchen, Germany) using a capillary die with L/D = 30/1, at the temperatures and shear rates indicated in Results and Discussion section. Filaments to be used for FFF 3D printing were prepared in a Haake MiniLab extruder (Thermo Fisher Scientific) at a temperature of 140 °C and diameter die of 1.75 mm. Obtained filaments presented a 1.70 mm ± 0.5 mm diameter and their compressibility modulus, K, was determined by compression stress-strain tests carried out in an Instron 5569.

The FFF printing process was performed in a Bowden type Voladora NX printer (Tumaker) using 0.4–1.2 mm nozzles and the printing geometries were originally designed in Solidworks 2016 x64 Editor and printed using Simplify printing program. Printing substrate temperature and printing velocity are detailed in the Results and Discussion section. Cohesive aspect of the printed objects was investigated by SEM (Hitachi S-2700) microphotographs of the welding.

3 Results and Discussion

Filament buckling occurs when low values of the parameter K/η are involved [1, 2]. On this basis, the results of Fig. 1 are not very encouraging, because the K/η values of our samples are indeed below those of reference polymers that do not show buckling.

Surprisingly enough, no buckling was actually observed during printing the filaments and Fig. 2 shows two specimens fabricated under the following conditions:

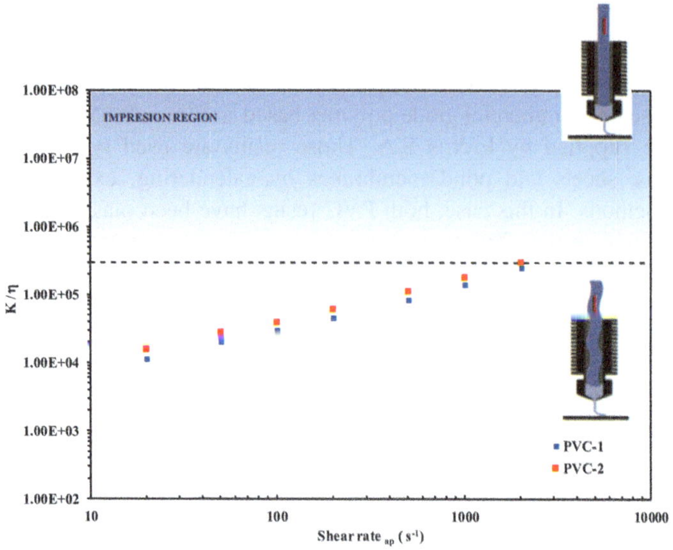

Fig. 1. *K/η* results for PVC-1 and PVC-2 at 180 °C. The dotted line delimited de empiric value (K/η = 3 10^5) determined by Venkarataman [1, 2].

nozzle T = 210 °C, nozzle diameter = 0.4 and 1.2 mm, printing velocity = 40 mm/s and 30% infill [3].

Fig. 2. Images of 3D specimens obtained with PVC-2 formulation (T = 210 °C, nozzle diameter = 0.4 and 1.2 mm, printing velocity = 40 mm/s)

The specimen presented good cohesive aspect and flexibility; as shown by SEM results of Fig. 3, suitable adhesion between consecutive layers is observed.

4 Conclusions

The remarked success in FFF 3D printing of plasticized PVCs and, in particular, the absence of buckling, constitutes an unexpected result, in view of the poor mechanical strength of the filaments and low *K/η* values. This disagreement between a negative

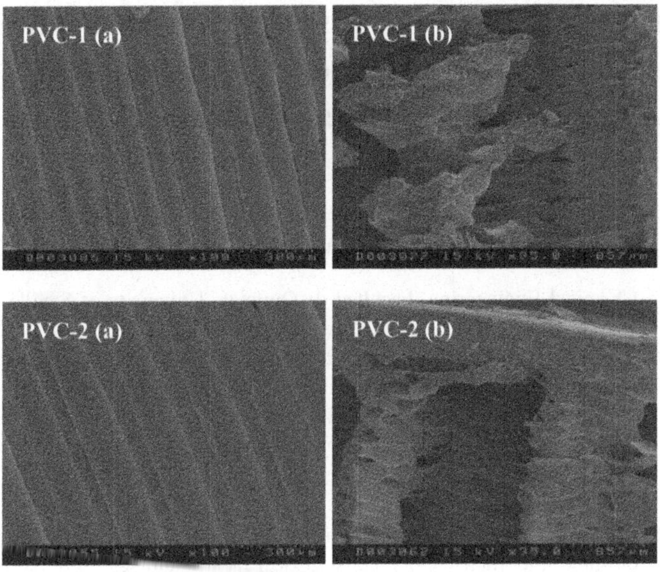

Fig. 3. SEM images of dumbbell shaped specimens obtained with PVC-1 and PVC-2 formulations: (a) external face and (b) cross-section.

foresight based on rheological assumptions and the reality of a good 3D printing, leads to reconsider our viscosity results, which have been obtained in a extrusion rheometer in temperature conditions not exactly similar to those of the printing device nozzle. Our hypothesis is that slippage or plug flow takes place during the extrusion of the plasticized PVC, reducing considerably the flow resistance and so avoiding buckling despite the low *K* values of the samples. Experimental studies on slippage in the nozzle are currently in progress in our laboratory.

Acknowledgments. This work was financially supported by ERCROS S.A. and by UPV/EHU (UFI 11/56) and GIC IT-586-13 (Basque Government).

References

1. Venkataraman, N., Rangarajan, S., Harper, B., Matthewson, M.J., Safari, A., Danforth, S.C.: Process-property-performance relationship for fused deposition of ceramics (FDC) feedstock materials. In: Danforth, S.C., Dimos, D., Prinz, F.B. (eds.) Solid Freeform and Additive Fabrication, San Francisco, CA, pp. 203–210 (2000)
2. Venkataraman, N., Rangarajan, S., Matthewson, M.J., Harper, B., Safari, A., Danforth, S.C., Wu, G., Langrana, N., Guceri, S., Yardimci, A.: Feedstock material property-process relationships in fused deposition of ceramics (FDC). Rapid Prototyping J. **6**, 244–252 (2000)
3. Peñas, M.I.: Reología e Impresión 3D. Nuevos materiales basados en PVC. Master thesis (2018)

Development of Porous Matrices as Scaffolds for Tissue Engineering: Rheological and Microstructural Characterization

Victor Perez-Puyana[1(✉)], Mercedes Jiménez-Rosado[1], Manuel Felix[1], Alberto Romero[2], and Antonio Guerrero[1]

[1] Chemical Engineering, Facultad de Química, University of Seville, 41012 Seville, Spain
{vperezll, mjimenez42, mfelix, aguerrero}@us.es
[2] Chemical Engineering, Facultad de Física, University of Seville, 41012 Seville, Spain
alromero@us.es

Abstract. The goal of tissue engineering is to repair damaged tissues. It is based on three important factors: cells, growth factors and scaffolds. The scaffolds are porous matrices which serve as a platform for cell adhesion and offer the ideal environment for the release of cells and so, their growth. In addition, the scaffold must keep its integrity until the tissue formed has suitable mechanical properties. Among the possible materials used for the development of scaffolds, two biopolymers can be highlighted: a protein such as collagen and a polysaccharide such as chitosan, which provide enormous biocompatibility. One of the challenges in these scaffolds is to provide the appropriate rheological properties during their use. To achieve it, an additional crosslinking process is carried out, to produce bonds between different biopolymer chains. These bonds can be produced in two different ways: chemical and enzymatic pathways. The results show that it is possible to develop scaffolds with enhanced mechanical properties and optimal porosity by modifying the initial composition used.

1 Introduction

Tissue Engineering (TE) requires structures that act as support for optimal cell growth. These structures are called scaffolds. Scaffolds are porous matrices that must provide enough mechanical support to withstand forces *in vivo* and maintain a potential space for tissue development. In addition, these scaffolds must keep their properties until the tissue formed has enough mechanical integrity and the cells express the appropriate genes to maintain the specific function of the tissues formed [1]. For this reason, tissue engineering is a multidisciplinary field that requires knowledge in chemistry, materials science and biology in order to analyze the different aspects of TE, with the aim of developing 3D matrices that offer the ideal environment for the release of cells and their initiating molecules, and thus, being able to potentially build biocompatible organs [2]. Furthermore, rheology is also an important field to consider in the development of these materials, since it is necessary to carry out a complete characterization of their structure

F. J. Galindo-Rosales et al. (Eds.): IBEREO 2019, SPM, pp. 148–151, 2020.
https://doi.org/10.1007/978-3-030-27701-7_31

to ensure their proper behavior. Moreover, these scaffolds are typically subjected to dynamic stresses inside the bioreactor in order to stimulate cell proliferation.

In this context, the aim of this work was to develop scaffolds with a combination of collagen (CG) and chitosan (CH) through a freeze-drying process. These materials, after processing, had to be rheologically and microstructurally characterized in order to verify that they fulfill the appropriate functions. Therefore, the mechanical behavior was analyzed by performing dynamic compression strain and frequency sweep tests and the microstructure was evaluated through SEM microscopy. In addition, enzymatic and chemical crosslinking agents were evaluated to strengthen the structure of the scaffolds.

2 Materials and Methods

The raw material used was type I collagen of porcine origin (CG), supplied by Essentia Protein Solutions (Denmark), and chitosan (CH), provided by Sigma Aldrich (USA). The solvent used is 0.05M acetic acid. Glutaraldehyde (Glut) and commercial enzyme transglutaminase (Tgase) have been used as chemical and enzymatic crosslinking agents, respectively. Both crosslinkers have been used with a concentration of 10 wt.% respect the polymer concentration.

2.1 Scaffolds Fabrication Technique

These biopolymer-based matrices (scaffolds) have been obtained through a 2-stage process, which consists in the freezing (−40 °C) of a previously prepared solution followed by a freeze-drying process during 24 h (−80 °C, < 15 Pa). Most scaffolds have been developed with the same protocol except in the case of adding transglutaminase as a crosslinking agent. This latter process undergoes a slight modification to ensure optimal conditions of action for the enzyme (50 °C). Scaffolds have been processed following a 1:1 collagen:chitosan formulation (1:1 CG:CH) at 1 wt%.

2.2 Scaffolds Characterization

Microstructural Characterization

Porosity: The porosity (ε) of the scaffolds was obtained through a method previously described [3], using the following equation:

$$\varepsilon(\%) = (1 - (\rho_s/\rho_m)) \cdot 100$$

where ρ_s is the density of the scaffold (calculated using the weight and volume of each scaffold) and ρ_m is the density of collagen type I, which is 0.68 $g{\cdot}cm^{-3}$.

Porosimetry: Low pressure mercury porosimetry was performed to investigate the mean pore size. This technique was performed using a PoreMaster-60 GT porosimeter (Quantachrome Instruments, USA).

Rheological Characterization
Dynamic compression tests were carried out using an RSA3 (TA Instruments, USA) rheometer, with a circular shaped geometry (dia: 15 mm) and a smooth surface. Two types of tests were conducted at 25 °C: (1) Strain sweep tests in a range between $2.5 \cdot 10^{-4}$ and 2.5% at 1 Hz. The linear viscoelastic range and the critical strain of the scaffolds were determined, considering the critical strain as the maximum strain allowed by the scaffold without breaking its microstructure; and (2) Frequency sweep tests in a range between 0.02 and 20 Hz, at a constant strain within the linear viscoelastic range of each system. In both tests, the elastic modulus (E') and viscous modulus (E'') were evaluated.

3 Results and Discussion

The data obtained on the porosity of the different elaborated systems are shown in Table 1. The percentage of porosity remains slightly stable, but the average pore size varies considerably according to the crosslinking method used, due to a microscopic restructuration of the scaffold.

Table 1. Porosity and mean pore size of 1:1 CG:CH scaffolds (reference) and 1:1 CG:CH scaffolds crosslinked by chemical (Glut) and enzymatic (Tgase) crosslinking agent.

Systems	Porosity (%)	Mean pore size (μm)
Reference	99.3	76.9
Chemical crosslinking (Glut)	98.9	50.1
Enzymatic crosslinking (Tgase)	99.4	38.6

The rheological properties are also important for the proper behavior of the scaffolds. The mechanical properties of the scaffolds produced can be observed in Fig. 1. All the systems have a predominantly elastic gel-like behavior since the storage modulus (E') is higher than the loss modulus (E''). As may be observed in Fig. 1A, crosslinking induces a reduction in the elastic modulus, particularly with Tgase.

In addition, Fig. 1B shows the evolution of the critical strain after performing the two different crosslinking methods. The use of both crosslinkers improved the critical strain of the system, highlighting the latter due to the improvement achieved. The results obtained show the possibility of producing a scaffold with a greater elastomeric character by the addition of a crosslinking stage during its fabrication process.

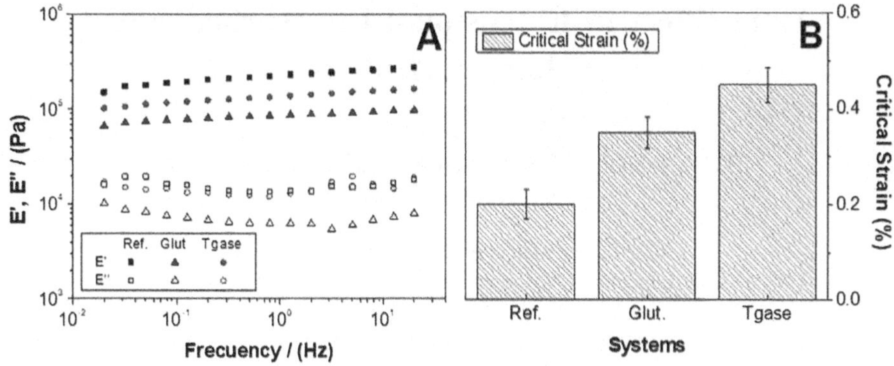

Fig. 1. (A) Frequency sweep tests and (B) critical strain values of 1:1 CG:CH scaffolds (reference) and 1:1 CG:CH scaffolds crosslinked by chemical (Glut) and enzymatic (Tgase) crosslinking agent.

4 Conclusions

This study demonstrated the possibility of developing scaffolds with 1% collagen (CG) and chitosan (CH) (1:1 CG:CH), with suitable mechanical and morphological properties. The addition of chemical or enzymatic crosslinking agents gives the scaffolds a greater elastomeric character, as reflected by the increase in critical strain. Although this increase leads to a moderate reduction in the elastic modulus and average pore size, the increase in the deformation may be beneficial when using a bioreactor.

Acknowledgements. This work is part of a research project (Ref. CTQ2015-71164-P) of MINECO/FEDER, EU. The authors gratefully acknowledge their financial support. In addition, authors also acknowledge for the pre-doctoral fellowships of Víctor Pérez Puyana (VPPI-US) and Mercedes Jiménez Rosado (FPU17/01718-MEFP), and the post-doctoral fellowship of Manuel Félix Ángel (VPPI-US).

References

1. Jana, S., Tefft, B.J., Spoon, D.B., Simari, R.D.: Scaffolds for tissue engineering of cardiac valves. Acta Biomater. **10**(7), 2877–2893 (2014)
2. Van Vlierberghe, S., Dubruel, P., Schacht, E.: Biopolymer-based Hydrogels as scaffolds for tissue engineering. Biomacromolecules **12**(5), 1387–1408 (2011)
3. Al-Munajjed, A.A., Hien, M., Kujat, R., Gleeson, J.P., Hammer, J.: Influence of pore size on tensile strength, permeability and porosity of hyaluronan-collagen scaffolds. J. Mater. Sci. Mater. Med. **19**, 2859–2864 (2008)

Time Effect in Extensional Electrorheological Characterization

H. H. Najafabadi[1], S. H. Sadek[1], Laura Campo-Deaño[1], and F. J. Galindo-Rosales[2(✉)]

[1] CEFT, Departamento de Engenharia Mecânica, Faculdade de Engenharia da Universidade do Porto, Rua Dr. Roberto Frias, 4200-465 Porto, Portugal
{hossein, campo}@fe.up.pt, sadek@gcloud.fe.up.pt
[2] CEFT, Departamento de Engenharia Química, Faculdade de Engenharia da Universidade do Porto, Rua Dr. Roberto Frias, 4200-465 Porto, Portugal
galindo@fe.up.pt

Abstract. When performing extensional ER tests for the characterization of electro-rheological fluids, the time between applying the electric field and triggering the experiment (delay time) plays an important role in ER behaviour of the sample. The current study investigates the delay time effect in extensional ER characterization. The ER sample used in the measurements is a suspension of cornstarch in olive oil with a concentration of 30 wt%. Fixtures adaptable to the commercial version of the Capillary Breakup Extensional Rheometer (Haake™ CaBER™, Thermo Scientific) are designed and fabricated to allow the application of an external electric field aligned with the flow kinematics undergone by the fluid sample while the extensional characterization is taking place. An *in-house* code developed in Matlab is used to process the images and investigate the filament thinning behaviour of the sample under 3 kV applied voltage and at different delay times (0 s, 10 s, and 40 s).

The results show that, under extensional flow, when the voltage is applied, and the delay time between setting the electric field and triggering the experiment is 10 s, the fluid filament breaks. However, if the delay time is increased up to 40 s, then it is observed that the filament does not break anymore. Thus, increasing the time between applying the voltage and triggering the experiment strengthens the structure of the ER sample. According to these results, it can be concluded that the formation of the microstructure due to the presence of the electric field does not occur instantaneously, but instead it is a time-dependent phenomenon.

1 Introduction

Electrorheological fluids (ERFs) are a type of smart fluids in which viscosity changes in response to an applied electric field. The response time of ERFs can be as short as a few milliseconds to switch from a liquid-like material to a solid-like material. Due to this marvelous feature, ERFs can be used as an electric-mechanical interface and can convert various mechanical devices such as dampers, clutches and valves into active mechanical

F. J. Galindo-Rosales et al. (Eds.): IBEREO 2019, SPM, pp. 152–156, 2020.
https://doi.org/10.1007/978-3-030-27701-7_32

elements. Since their discovery six decades ago by Winslow [1], ERFs have attracted a lot of scientific curiosity, due to their diverse application potential [2, 3].

A generally accepted mechanism is electrostatic polarization, which attributes the origin of electrorheological effect to the polarization of the dispersed phase particles relative to the continuous phase in the presence of an electrical field. In this model, the particles form fibrillated chains, which cause abrupt modification of the rheological properties.

The rheological properties of ERFs should be measured in various canonical flows under the presence of an electric field to attain a complete description and better understanding of these fluids. Many conventional rotational rheometers are adequately modified to allow the application of an electrical field on the sample. However, the measurement of rheological properties in uniaxial extensional flows is still relatively new and there are no studies about the application of the electrical field for extensional measurements. Sadek et al. [4], fabricated a novel prototype consisting of unique fixtures to apply the electric field on the sample fluid in parallel with the extensional flow applied by the CaBER device. This prototype is used in the current study for the application of electrical field on the CaBER device.

The time between applying the electric field and triggering the rheological experiment, hereafter named as delay time, should be thoroughly investigated in the extensional electrorheological characterization of the fluid. In this work, the effect of delay time on extensional electrorheological properties of a concentrated suspension of corn starch/olive oil is studied.

2 Materials and Methods

An ER Fixture adaptable to the CaBER device (Fig. 1) recently developed by Sadek et al. [4], allows the application of an electric field on the sample fluid aligned with the extensional flow. An external electrical power source generates and controls the electric field.

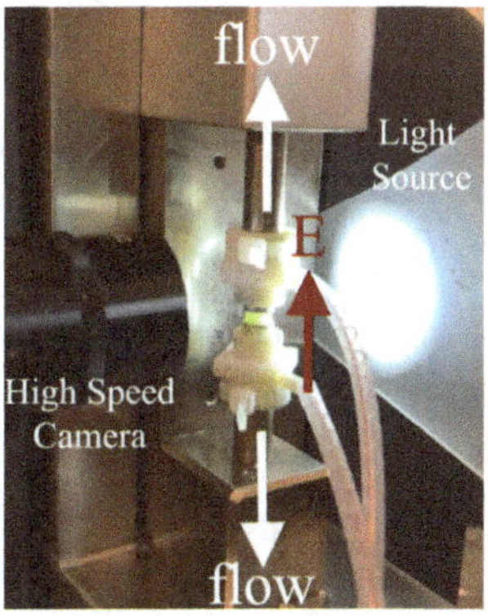

Fig. 1. ER Fixture adaptable to the CaBER device developed by Sadek et al. [4].

For this study, 4 mm diameter plates were selected and the initial gap between the plates was set at 2 mm. The final axial separation between the plates was fixed at 8 mm and the initial stretch profile was linear with a "strike time" of 20 ms. The filament thinning process was recorded from the left side with a high-speed camera (Photron FASTCAM Mini UX100) at 2000 fps, as in [5]. The time evolution of the filament was obtained from the set of images using the Matlab Image Processing Toolbox. The rheological behavior of cornstarch/olive oil suspension is characterized under extensional flow at a concentration of 30 wt% applying a voltage difference of 3 kV. Fresh samples are used for each measurement after being re-dispersed in an ultrasound bath for 10 min. The time evolution of the filament mid-diameter at the applied voltage and delay times of 0 s, 10 s and 40 s were investigated.

Additionally, a rotational rheometer (Anton Paar MCR301) equipped with an electrorheological device and concentric cylinders measuring system was used for the analysis of the delay times (0 s and 40 s) on the electrorheological behavior under shear flow. Start-up experiments at a shear rate of 100 s^{-1} were carried out.

3 Results and Discussion

Figure 2 shows the time evolution of the shear viscosity at different delay times. A slight influence of the time delay on the transient behavior of the samples under shear flow is observed, resulting in lower viscosity values for larger delay times.

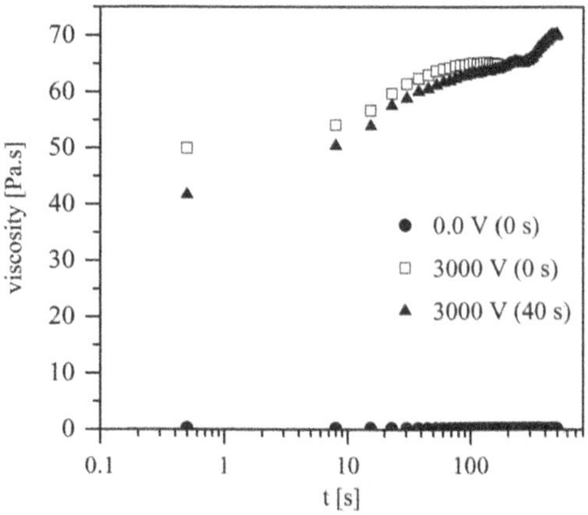

Fig. 2. Effect of delay-time on the time evolution of the shear viscosity under start-up experiments at $\dot{\gamma} = 100\ s^{-1}$, using 30 wt% cornstarch/olive oil at 0.0 V and 3 kV.

After ~100 s, all the transient responses tend towards the same trend. It is remarkable that the fact of increasing the delay time between the application of the field and the beginning of the startup experiments, results in a decrease in the initial shear viscosity. It has to be considered that under shear flow, the electrorheological cell applies the electric field perpendicular to the flow direction and so do the particle chains. The longer the delay time, the more the time to build the structure of chains, which results in a more brittle behavior under shear forces and, consequently, in a decrease in the measured shear viscosity.

Figure 3 shows the time evolution of the filament mid-diameter as a function of different delay times in the CaBER experiment. It is observed that under extensional flow and the applied voltage, the filament breaks when the delay time between setting the electric field and triggering the experiment is 10 s. However, the fluid filament does not break anymore if the delay time is increased up to 40 s. Thus, the structure of the ER sample is strengthened when the delay time is increased. Here the particle chains are aligned with the flow direction and help to extend longer the life of the filament. That is, the microstructure formation in the sample due to the presence of the electric field is not an instantaneous phenomenon, but instead it is clearly time-dependent and it is better sensed under extensional flows.

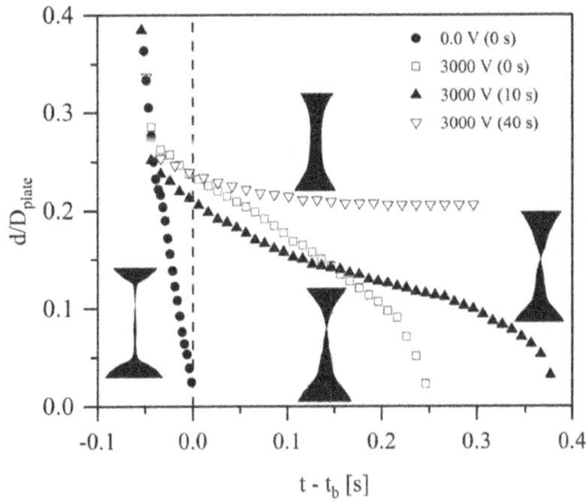

Fig. 3. Effect of delay-time on the time evolution of the mid-diameter for a 30 wt% cornstarch/olive oil suspension under extensional flow.

Acknowledgements. The authors acknowledge the financial support provided by Fundação para a Ciência e a Tecnologia (FCT), COMPETE and FEDER through projects MIT-EXPL/IRA/0077/2017 and POCI-01-0145-FEDER-030765.

References

1. Winslow, W.M.: Induced fibration of suspensions. J. Appl. Phys. **20**(12), 1137–1140 (1949)
2. Yin, J., Zhao, X.: Titanate nano-whisker electrorheological fluid with high suspended stability and ER activity. Nanotechnology **17**(1), 192 (2005)
3. Sheng, P., Wen, W.: Electrorheological fluids: mechanisms, dynamics, and microfluidics applications. Annu. Rev. Fluid Mech. **44**, 143–174 (2012)

4. Sadek, S.H., Najafabadi, H.H., Campo-Deaño, L., Galindo-Rosales, F.J.: Extensional Electro-Rheological Fixture (ExERF). Provisional Patent Application Ref. PAT20191000021255/0198 (2019)
5. Campo-Deaño, L., Clasen, C.: The slow retraction method (SRM) for the determination of ultra-short relaxation times in capillary breakup extensional rheometry experiments. J. Non-Newton. Fluid Mech. **165**, 1688 (2010)

Non-Newtonian Fluid Mechanics

Effect of a Constant Drift in the Reptation Dynamics of Entangled Polymers

Andrés R. Tejedor and Jorge Ramírez(✉)

Department of Chemical Engineering, Universidad Politécnica de Madrid, Madrid, Spain
jorge.ramirez@upm.es

Abstract. In this work, we explore a modified reptation theory for entangled polymers which includes an additional relaxation mechanism that allows the chain to escape the tube: a constant drift velocity along the primitive path. It is shown that, if the drift is sufficiently small, the Gaussian hypothesis of the tube theory is respected and, if it is larger than a molecular weight-dependent threshold, it can dominate the dynamics at long times, yielding interesting new behaviour.

1 Reptation with Drift

The most successful theory to describe the dynamical response of entangled polymer melts is the reptation/tube theory of de Gennes and Edwards [1,2]. In the basic version of the theory, a single tracer polymer chain moves through the matrix created by surrounding polymers, which create topological restrictions (named entanglements) due to the uncrossability of the chains. This mesh of obstacles is replaced by a mean field representation (a tube of diameter a), and the tracer chain is free to diffuse along the axis of the tube (which is called the primitive path). This diffusion motion (called reptation by de Gennes) constitutes the main relaxation mechanism of well entangled linear polymer chains. The original model has been modified and updated thoroughly in the past decades with many additional relaxation modes (contour length fluctuations [3], constraint release [4,5], tube dilution [6,7], convective constraint release [8]) that allow the model to explain experimental data of binary blends and branched polymers in linear and non-linear flow conditions.

In addition, the tube theory has been useful to describe the diffusion of charged biomolecules in electrophoresis experiments, in the form of the biased-reptation theory [9]. In the experiments, a long, charged polymer diffuses through a gel under the effect of an external electric field, which enforces the chain to reptate preferentially in the direction of the electric field and tends to align the end-to-end vector with the field.

Here, we consider a similar but different setup: we envision a linear, flexible polymer chain which reptates with a constant drift velocity always pointing to

F. J. Galindo-Rosales et al. (Eds.): IBEREO 2019, SPM, pp. 159–162, 2020.
https://doi.org/10.1007/978-3-030-27701-7_33

the same direction along the tube (see Fig. 1). This propulsion can be due to some internal activity or to the effect of a conformational asymmetry along the chain contour. The velocity c is limited to assure the random walk statistics of the primitive path. When this kind of drift is introduced in the reptation model of de Gennes, the dynamical behaviour of the system (e.g. the tube segment survival function, the segmental motion, the dynamic structure factor, etc) changes dramatically. In this work, we explore these changes by means of analytical theory and simulations.

2 Dynamic Behaviour of a Polymer with Drift

The main hypothesis in our theory is that there is a constant drift that always points in the same direction along the tube and that is fast enough to affect diffusion at long times, but slow enough so that the tube ends are free to explore all possible orientations when they exit the tube (see Fig. 1). Within this well-defined framework, we have analytically solved the partial differential equations of reptation with drift for the tube survival function, the segmental motion and the dynamic structure factor, and compared them to the solutions found in the seminal work of Doi and Edwards in the case of pure reptation [2].

Our results show substantial differences with respect with the original reptation model. Non-surprisingly, we have observed that diffusion governs the molecular motion at very small times (displacement scales with $t^{1/2}$) whereas drift may govern at long times (motion scales with t), if the drift velocity is larger than a critical value that scales with the inverse of the squared number of entanglements Z^{-2}. Therefore, no matter how small the drift is, there is always a sufficiently large number Z that makes the drift govern the dynamics over reptation in the terminal region.

This effect can be appreciated by studying several observable functions. For instance, in Fig. 2 we present the tube segment survival function $\psi(s,t)$ measured at different times for a chain with $Z = 32$ entanglements and two different drift

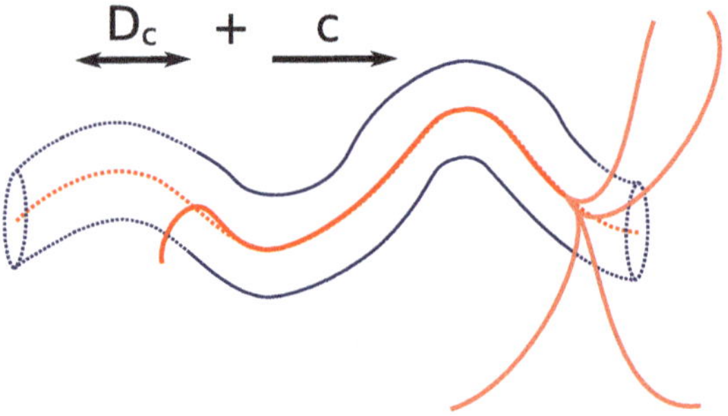

Fig. 1. Scheme of a chain escaping out of its confining tube with the help of reptation with diffusion coefficient D_c and a drift velocity c. The drift drives the polymer in a preferential direction along the primitive path.

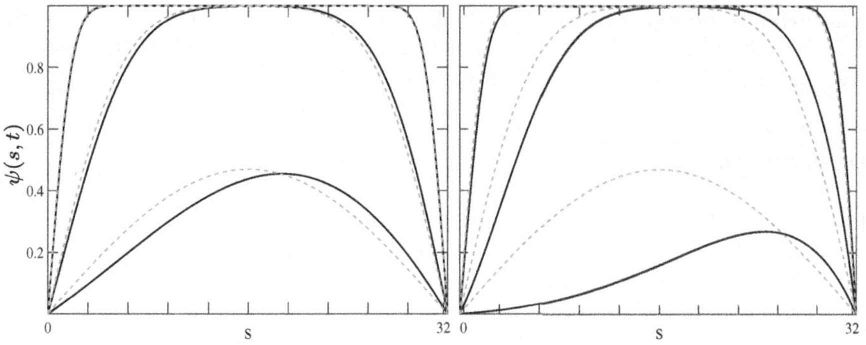

Fig. 2. Tube segment survival function for pure reptation (dashed lines) and reptation with drift $c = 6.1 \times 10^{-5}$ (left) and 0.063 (right) in solid lines, computed at times $t = (0.01, 0.1, 1) \times \tau_d$.

values. It is remarkable that a small drift does not seem to affect the behavior of the original tube survival function significantly (left panel), whereas it has a strong influence over the dynamics of the chain when the drift is large enough (right panel). It is important to remind that, in the tube theory, relaxation modulus and viscosity can be extracted from $\psi(s, t)$. In the limit of strong drifts, the dependence of the viscosity with the molecular weight changes dramatically, becoming linear instead of the cubic power law of pure reptation. Similarly, the segmental motion and center of mass diffusion can be determined analytically and, in the limit of significant drift velocity, the self-diffusion coefficient becomes independent of the molecular weight of the polymer. From our analytical study, we have inferred a molecular weight-dependent range of velocities for which the drift mechanism governs over reptation.

All our analytical results have been verified by means of Brownian dynamics simulations of a discrete version of the tube model (a 1D Rouse model diffusing inside a 3D random walk) considering contour length fluctuations (CLF) and without constraint release.

3 Conclusions

In this work, we have studied in detail the response of entangled polymer melts when they are subjected to an additional mechanism of drift, and we have compared the most relevant features of this model with the original work of Doi and Edwards [1,2]. The drift reduces the time needed by the chain to escape from the tube, and this reduces the viscosity and increases the diffusion coefficient significantly. This behaviour could be very relevant if the kind of drift motion that we hypothesize in our work is found in real systems. For example it will allow to create very high molecular weight polymeric materials with very low viscosity and high diffusion coefficient. The full details and analytical calculations of this work will be published in a forthcoming article [10].

Acknowledgements. ART and JR acknowledge funding from grant PEJ-2017-AI/IND-6767 of the Regional Government of Madrid and project UPM RP 160543006 from Universidad Politécnica de Madrid.

References

1. de Gennes, P.G.: J. Chem. Phys. **55**, 572 (1971)
2. Doi, M., Edwards, S.: The theory of polymer dynamics **73**, 190–205 (1988)
3. Doi, M.: J. Polymer Sci.: Polym. Phys. Edn. **21**, 667–684 (1983)
4. des Cloizeaux, J.: J. de Phys. I **3**, 61–68 (1993)
5. Rubinstein, M., Colby, R.H.: J. Chem. Phys. **89**, 5291 (1988)
6. Marrucci, G.: J. Polym. Sci.: Polym. Phys. Edn. **23**, 159–177 (1985)
7. Milner, S.T., McLeish, T.C.: Macromolecules **30**, 2159–2166 (1997)
8. Graham, R.S., Likthman, A.E., McLeish, T.C., Milner, S.T.: J. Rheo. **47**, 1171 (2003)
9. Viovy, J.L.: Rev. Mod. Phys. **72**(3), 813 (2000)
10. Tejedor, A.R., Ramirez, J.: in preparation (2019)

Suspensions and Colloids

Rheology and Physical Stability of Rosemary Essential Oil Emulsions

María José Martín-Piñero, Jenifer Santos García(✉),
Luis Alfonso Trujillo-Cayado, María Carmen García González,
and Maria Carmen Afaro Rodríguez

Departamento de Ingeniería Química, Facultad de Química,
Universidad de Sevilla, Seville, Spain
{mjmartin, jsantosgarcia, ltrujillo,
mcgarcia, alfaro}@us.es

Abstract. The main objective of this work was to study the influence of the welan gum and advanced performance xanthan gum concentration on rheological properties and physical stability of rosemary essential oil emulsions. The increase of biopolymer concentration provoked an increase of complex viscosity. Concerning the physical stability, for the higher concentration of gum, creaming was eliminated but flocculation occurred. This flocculation led to destabilization by coalescence, being the emulsion formulated with 0.4 wt% the most stable.

1 Introduction

In this work, the rheology and physical stability of rosemary essential oil emulsions stabilized with welan gum (WG) and advanced performance xanthan gum (APXG) were investigated and compared. Both gums are anionic polysaccharides obtained by aerobic fermentation from *Sphingomonas sp.* ATCC 31555 and *Xanthomonas campestris,* respectively [1]. These gums find applications in different fields such as in the food, cosmetic or agrochemical industries. They are often used to increase the viscosity of the continuous phase and to avoid or slow down some destabilization processes, like sedimentation and creaming. However, other studies reported that low concentration of biopolymer promotes creaming due to depletion flocculation. Increasing the concentration of gum, a three-dimensional gel-like network of droplets in emulsion is formed due to the flocculation of droplets [2]. This study assesses the influence of WG and APXG concentration on rheological properties and physical stability of O/W emulsions formulated with rosemary essential oil. In order to know the linear viscoelastic properties and the flow behaviour of emulsions, small amplitude oscillatory shear (SAOS) experiments and flow curves were carried out, respectively. In addition, multiple light scattering was used to determine the physical stability for all emulsions studied.

F. J. Galindo-Rosales et al. (Eds.): IBEREO 2019, SPM, pp. 165–169, 2020.
https://doi.org/10.1007/978-3-030-27701-7_34

2 Materials and Methods

2.1 Materials

Rosemary essential oil was supplied by Sigma Aldrich and Appyclean 6548 was purchased from Wheatoleo. Additionally, xanthan gum (KELTROL® Advanced Performance), and welan gum (K1A96) were kindly provided by CP Kelco Company (San Diego, USA). Sodium azide (0.1 wt% in the final formulation) was added to the samples to prevent the growth of microorganisms. Milli-Q water was used to prepare the aqueous phase.

2.2 Emulsions Developed

All emulsions were prepared according to the emulsification method previously reported [3]. The concentration of oil and emulsifier was fixed at 20 wt% and 4 wt%, respectively. Finally, the gum was added and it was stirred using an IKA for two hours up to a maximum of 800 rpm.

2.3 Physical Stability

The physical stability of rosemary essential oil emulsions with gums were evaluated for 28 days at 30 °C by a Turbiscan Lab Expert device (Formulaction, France) using the multiple light scattering technique. In order to evaluate the stability, the parameter TSI (Turbiscan Stability Index) was calculated by the Turbiscan software using the following formulae:

$$TSI = \sum_{i} \frac{\sum_{h} |scan_i(h) - scan_{i-1}(h)|}{H} \quad (1)$$

where $scan_i$ is the average backscattering for each time (i) of measurement, $scan_{i-1}$ (h) is the average backscattering for the (i − 1) time of measurement and H is the number of scans carried out on the sample.

2.4 Rheology

Flow curves were performed with a CS Haake-MARS rheometer (Thermo) and a double cone geometry (60 mm/1°) for emulsions with gum concentration lower than 0.3 wt%. Emulsions with 0.4 and 0.5 wt% of gums were measured using a serrated plate-plate sensor of 60 mm of diameter. Flow curves were carried out in the 0.05–20 Pa shear stress range using a step-wise method.

The mechanical spectra were obtained from 0.05 to 15 rad/s at fixed shear stress within linear viscoelastic region previously obtained by stress sweeps. All measurements were done by triplicate and the values shown are the average of the three replicates. The standard deviation of all measurements was lower than 5–10% for all the samples.

3 Results and Discussion

Zero-shear viscosity values (η_0) at one day of aging time are shown in Fig. 1A as a function of gum concentration for either gum. All emulsions presented a high shear-thinning behavior. Thus, zero-shear viscosity was calculated by fitting the flow curves fairly well to the Cross model ($R^2 > 0.99$):

$$\eta = \frac{\eta_0}{1 + \left(\frac{\dot{\gamma}}{\dot{\gamma}_c}\right)^{1-n}} \quad (2)$$

where $\dot{\gamma}_c$ is related to the critical shear rate for the onset of the shear-thinning response, η_0 stands for the zero-shear viscosity and $(1 - n)$ is a parameter related to the slope of the power-law region; n being the so-called "flow index".

As expected, η_0 increases as a function of gum concentration. This fact is due to not only the thickening effect of polysaccharides but also may be due a depletion flocculation induced by the presence of gum in the continuous phase. At 28 days of aging time (data not shown), an increases of η_0 occurred at low concentration of gum. An increase of zero-shear viscosity is related to a destabilization process by creaming. By contrast, zero-shear viscosity decreased for 0.3, 0.4 and 0.5 wt% of gum. This decrease in η_0 with aging time indicated a coalescence destabilization mechanism. Figure 1B shows the influence of the frequency on the complex viscosity of emulsions studied at several concentrations and types of gum. A clear increase of the viscoelasticity of emulsions studied can be observed by increasing the gum concentration, regardless of gum type used. The emulsions formulated with APXG exhibited higher |η*| values than that one formulated with WG, although the differences observed decrease as increasing the concentration.

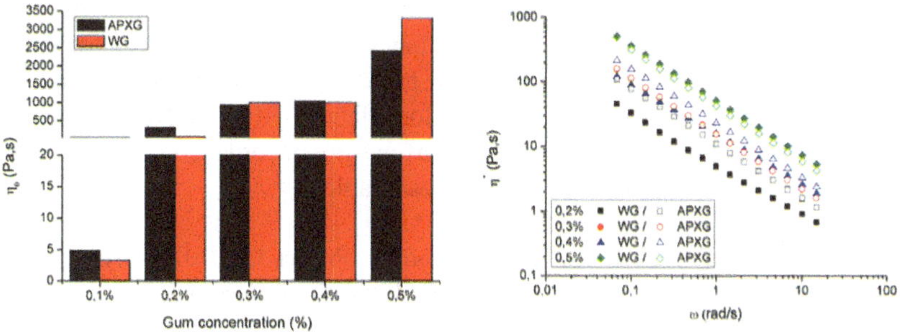

Fig. 1. (A) Zero-shear viscosity versus concentration of gum used in the formulation of rosemary oil/W emulsions aged for 24 h. (B) Influence of frequency on the complex viscosity of rosemary oil/W emulsions formulated with WG and APXG at several concentrations. T = 20 °C

Multiple light scattering revealed the occurrence of a destabilization by creaming for emulsions formulated with a low concentration of gum while the emulsions formulated with a gum concentration higher than 0.3 wt% were destabilized by flocculation which led to coalescence. These destabilization processes are in accordance with o the results previously obtained. The Global TSI values for all emulsions studied are shown in Fig. 2. It can be observed that TSI was lower (higher stability) for those samples formulated with 0.4 wt% of gum, regardless of the gum type. Additionally, it can be observed slight higher stability for emulsions formulated with APXG.

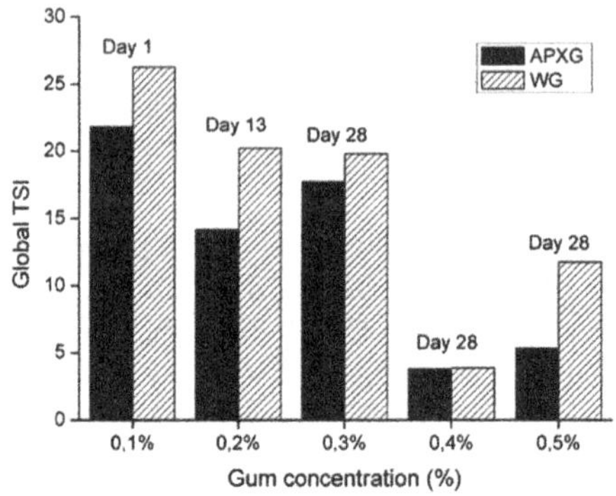

Fig. 2. Global TSI values for all studied emulsions aged for 24 h (0.1 wt%), 13 days (0.2 wt%) and 28 days (0.3 wt%, 0.4 wt% and 0.5 wt%)

4 Conclusions

Stable rosemary essential oil emulsions were developed by microfluidization technique. The addition both of WG and APXG provokes an increase of viscosity. Both gums showed shear-thinning behaviour and were fitted to the Cross model. η_0 and $|\eta^*|$ parameters gave information about not only the mean destabilization process, such as coalescence or creaming, but also on the flocculation grade. Multiple light scattering illustrated that creaming was the most important destabilization process for emulsions with 0.1 and 0.2 wt% of WG and APXG. At higher gum concentration, creaming was eliminated but flocculation occurred. This flocculation led to destabilization by coalescence. All these destabilization mechanisms are in concordance with rheology. The emulsion formulated with 0.4 wt% is the most stable.

Acknowledgements. The financial support received (Project CTQ2015-70700-P) from the Spanish Ministerio de Economia y Competitividad and the European Commission (FEDER Programme) is kindly acknowledged.

References

1. Xu, L., Xu, G., Liu, T., Chen, Y., Gong, H.: The comparison of rheological properties of aqueous welan gum and xanthan gum solutions. Carbohydr. Polym. **92**, 516–522 (2013)
2. Krstonošić, V., Dokić, L., Nikolić, I., Milanović, M.: Influence of xanthan gum on oil-in-water emulsion characteristics stabilized by OSA starch. Food Hydrocoll. **45**, 9–17 (2015)
3. Martin-Piñero, M.J., Ramirez, P., Muñoz, J., Alfaro, M.C.: Development of rosemary essential oil nanoemulsions using a wheat biomass-derived surfactant. Colloids Surf. B Biointerfaces **173**, 486–492 (2019)

Yield Stress in Injection Grouts for Strengthening of Stone Masonry Walls

Luis G. Baltazar[1(✉)], Fernando M. A. Henriques[1], Diogo Reis[1], and Maria Teresa Cidade[2]

[1] Department of Civil Engineering, Faculty of Sciences and Technology, NOVA University of Lisbon, Campus Caparica, 2829-516 Almada, Portugal
{luis.baltazar, fh}@fct.unl.pt,
dv.reis@campus.fct.unl.pt

[2] Department of Materials Science and CENIMAT|i3N, Faculty of Sciences and Technology, NOVA University of Lisbon, Campus Caparica, 2829-516 Almada, Portugal
mtc@fct.unl.pt

Abstract. Stone masonry is a simple and durable constructive technique that was exhaustively used until the mid 20th century. However, stone masonry has particular weaknesses which, associated with the absence of maintenance, increase the vulnerability of its structural integrity. For this reason, masonry walls often need consolidation to improve mechanical performance, such as adhesion between elements and load bearing capacity. Grout injection is a frequently used technique for the consolidation and strengthening of old stone masonry walls. The grout can be seen as suspension of binder particles in aqueous media that has a considerable fluidity in order to be pumped into voids and cracks within the masonry. For this reason, the rheology appears as a very useful tool in the design and quality control of the injection grout. The rheological behavior of hydraulic grouts for masonry consolidation is complex and some rheological properties (such as yield stress) are problematic to determine because of the combined effect of the hydration reactions of the binder and the interactions between the particles, present in the suspension. Despite the relevance of yield stress for injection grouts, no standard protocols are available and, therefore, the yield stress is often determined as an isolated parameter without taking into account phenomena such as thixotropy and hydration. In this study, the determination of yield stress of natural hydraulic lime-based grouts was performed with different measurements techniques using a rotational rheometer. The change in yield stress with time due to hydration was determined. Two yield stress, static and dynamic, and the critical shear rate were also measured, which can be used in the grouts design in order to achieve better grouting operation.

1 Introduction

Stone masonry is a simple constructive technique that has particular weaknesses which, associated with the absence of maintenance, increase the vulnerability of its structural integrity. Grout injection technique (or grouting) is often used as solution to overcome

F. J. Galindo-Rosales et al. (Eds.): IBEREO 2019, SPM, pp. 170–174, 2020.
https://doi.org/10.1007/978-3-030-27701-7_35

the structural problems of these old stone masonry walls by restoring the bonds between the masonry elements and, therefore, improving the load bearing capacity. Grouting consists of a suspension of binder particles in an aqueous medium that has a considerable fluidity in order to be pumped into voids within masonry. For this reason, the rheology appears as a very useful tool in the design and quality control of the injection grout [1]. Cementitious-based grouts are known as having a complex rheology. The yield stress of a grout will affect the relationship between injection pressure and flow. Håkansson [2] proposed two types of yield stress; the static yield stress that can be seen as the yield stress after the material leaves the rest, and the dynamic yield stress that has been defined as the yield stress when the material is under shearing i.e. in a broken-down state. The purpose of this study is to measure the yield stress of natural hydraulic lime-based grout, including the effect of thixotropy and hydration, by means of different measurement protocols.

2 Materials and Methods

2.1 Materials

Natural hydraulic lime (NHL5) was chosen because it is the hydraulic binder that presents properties closer to those of pre-existing materials in old masonries. A commercially available powder high range water reducer (HRWR), namely a polycarboxylate ether was used. All the tests in this work were based on NHL grout with a water to NHL ratio (w/b) of 0.4 and with 0.2 wt% of HRWR. The grout mixtures were prepared in laboratory in batches of 300 ml and the components mixed using a mechanical shear mixer.

2.2 Rheological Measurements

The rheological measurements were performed with a Bohlin Gemini HR^{nano} rotational rheometer (Malvern, UK). The parallel-plate geometry was used to perform all the measurements. The diameter of the geometry was 40 mm and the gap was 2 mm. The surface roughness of the upper plate was modified by means of an emery paper (grid 600) to minimize the slippage during the measurements. Controlled shear stress (CSS) was used to perform the stress ramp tests. A ramp of 0.3 Pa/s was applied and the stress was increased from 0.006 to 140 Pa; this was followed by a down ramp where the applied stress was decreased from 140 to 0.006 Pa. The corresponding shear rate and apparent viscosity were measured. The change with time of the static and dynamic yield stress was measured for 50 min. Moreover, creep tests were performed to examine the critical shear stress. To do so, four different stresses 0.006, 0.06, 0.6 and 6 Pa, were applied during 20 min. In addition, controlled shear rate (CSR) was also performed. The measurements were made in the shear rate range of 0.5–300 s^{-1} followed by a downwards curve in order to evaluate the existence of thixotropy. The change in the shear stress with time was also measured for 50 min.

3 Results and Discussion

3.1 Thixotropy and Hydration by CSR

A simple yield stress fluid typically does not depend on the shear history, which means that when solicited with a shear rate ramp the shear stresses overlap for the up and down curves. In contrast, for the case of NHL-based grouts, the yield stress will depend on the shear history, hydration process and, therefore, the measuring protocol. In this context, a series of controlled shear rate measurements were performed on the same sample during 50 min with 10 min interval between each measurement, as shown in Fig. 1.

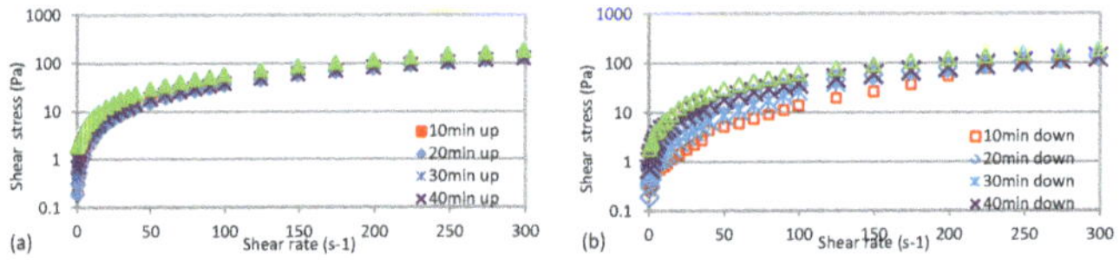

Fig. 1. Thixotropic behavior of NHL grout determined by controlled shear rate (a) up curves; (b) down curves

The effect of NHL hydration can be seen in Fig. 1, where the shear stress consistently increases with time. Moreover, it can also be seen that the up curve presents higher shear stresses than the down curve, which is a characteristic of thixotropic material. It should be highlighted the fact that thixotropy is less pronounced with time; this may be due to the hydration reactions that form bonds that are not destroyed by the application of the shear rate.

3.2 Yield Stress Obtained Through CSS

To determine the static and dynamic yield stresses of NHL grout, CSS measurements were performed because of their wide-ranging suitability for the determination of yield stress. As previously mentioned the grout samples were subjected to a CSS from 0.006 to 140 Pa for up curve and down curve. As shown in Fig. 2, the change in the apparent viscosity in the up curve before yielding and after yielding is rather sudden; i.e., the apparent viscosity tends to infinity at low shear stresses and there is a finite apparent viscosity at higher shear stresses.

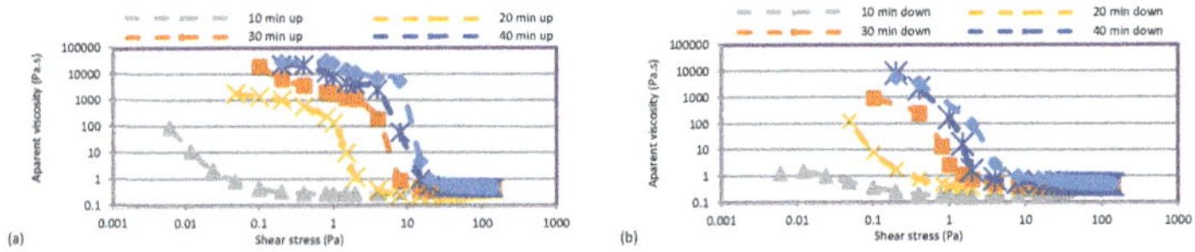

Fig. 2. Curves obtained from shear stress ramp: (a) up curves; (b) down curves

Based on the results presented in Fig. 2, it can be noted that the up curve showed a higher apparent viscosity and yield stress than the down curve due to the thixotropy. The yield stresses increase with time from 0.006 to 8 Pa. The yield stress increased to 1 Pa after 20 min and between 20 and 40 min the yield value increase from 2 to 4 Pa. Concerning the down curve, the yield stress increased with time, although with a slightly lower significance.

3.3 Yield Stress Obtained by Creep Test

The yield stress of colloidal suspensions can be determined through the creep test by detecting an abrupt change of apparent viscosity, i.e. the so-called viscosity bifurcation behavior. The variation in apparent viscosity for some applied stresses can be seen in Fig. 3. For an imposed stress of 0.006 Pa and 0.06 Pa, the apparent viscosity increased with time. The apparent viscosity below the yield stress would be infinite. In contrast, for the stress of 0.6 Pa the apparent viscosity decreases slightly and for the stress of 6 Pa the viscosity remains practically constant. Considering the results obtained, there remains a critical stress between 0.06 Pa and 0.6 Pa values that settles whether the grout would no flow due to build up or starts to flow due to the breakdown of the inter-particle bonds. It should be noted, however, that lower apparent viscosity values were obtained for the stress of 0.006 Pa than for the 0.06 Pa which can be explained by some slipping when such low shear stresses are imposed. Moreover, for the shear stress of 0.6 Pa a slight increase in viscosity at the end can be seen, such behavior can indicate that there is a competition between the micro-structure destruction and hydration reactions.

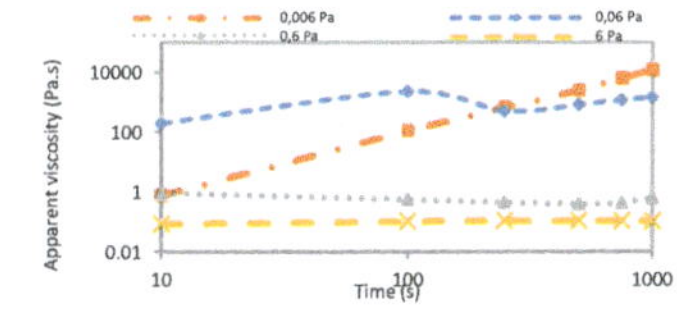

Fig. 3. Evolution of apparent viscosity as function of time and stress determined by creep test

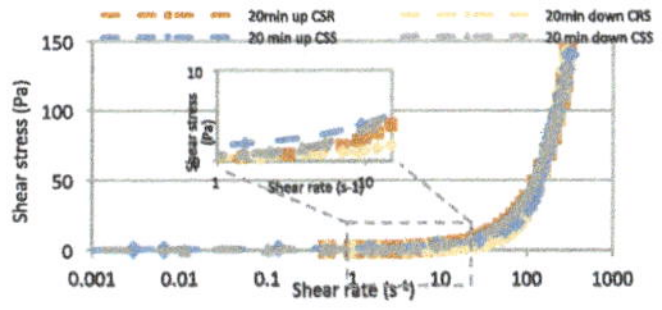

Fig. 4. Flow curves obtained by CSS and CSR measurement

3.4 Static and Dynamic Yield Stress

The results of CSS and CSR measurements at 20 min are shown in Fig. 4 (the other measurement periods presented similar behavior). It can be observed that regardless of the measurement performed the flow curves have a close behavior.

A slight linear increase in shear stress can be observed until 1 s^{-1}, for both up and down curves. A substantial increase in shear stress was seen for shear rates above 10 s^{-1}, which can justify the change of apparent viscosity observed in Fig. 2. This shear rate range between 1–10 s^{-1} can be considered as the transition zone between two yield stresses, which means that a shear rate until 1 s^{-1} would lead the static yield stress while shear rate higher than 10 s^{-1} would provide the dynamic yield stress. From a practical point of view, the yield stress must be chosen based on the typical shear rate range that the grout is subjected to. Moreover, the dynamic yield stress should be considered at the beginning of grouting when the grout is in a fully broken-down state. On the other hand, the static yield stress should be used as design parameter just for lower shear rates (i.e. at later stages of grouting operation) when the bonds between NHL particles starts to take place.

4 Conclusions

This study showed that the yielding values depend on the thixotropic behavior of NHL grouts and that two ranges of yield stress, static and dynamic, can be measured. The static yield stress was found to be in the range of 0.06–8 Pa and the dynamic yield stress in the range of 0.06–0.2 Pa. From the creep test, the grout showed a bifurcation behavior, and the critical shear stress was found to be between 0.06 Pa and 0.6 Pa. Moreover, it appears as the critical shear rate range, below which there is a transition from the dynamic to the static yield stress is in the shear rate range of 10–1 s^{-1}. These results indicate that at an early stage of the injection process the dynamic yield stress should be considered as design input whereas at a later stage, when the shear rate is slowing down, the static yield stress should be used.

References

1. Baltazar, L.G., Henriques, F.M.A., Cidade, M.T.: Rheology of natural hydraulic lime grouts for conservation of stone masonry—Influence of compositional and processing parameters. Fluids **4**(1), 13 (2019)
2. Håkansson, U.: Rheology of fresh cement based grouts (Ph.D. Thesis). Royal Institute of Technology, Stockholm (1993)

Microfluidics and Microrheology

Pressure Tap Influence on the Flow of Viscoelastic Fluids in a Microfluidic Channel

Tomás Rodrigues[1], J. Hermenegildo García-Ortiz[3], Francisco J. Galindo-Rosales[2], and Laura Campo-Deaño[1](✉)

[1] CEFT, Departamento de Engenharia Mecânica, Faculdade de Engenharia da Universidade do Porto, Rua Dr. Roberto Frias, 4200-465 Porto, Portugal
{tomas.rodrigues,campo}@fe.up.pt

[2] CEFT, Departamento de Engenharia Química, Faculdade de Engenharia da Universidade do Porto, Rua Dr. Roberto Frias, 4200-465 Porto, Portugal
galindo@fe.up.pt

[3] Departamento de Ingeniería Mecánica y Diseño Industrial, Escuela Superior de Ingeniería, Universidad de Cádiz, Cádiz, Spain
mere.garcia@uca.es

Abstract. How does flow inside a microfluidic device behave when passing by a pressure tap? Will its natural state be disrupted by any means? We seek to answer these questions and assess the consequences (if any) of attaching pressure taps to microchannels. To do so, streakline photography was performed for flow patterns visualisation in the close proximity of different types of pressure taps, using 'long' exposure times. Both a Newtonian and a viscoelastic fluid were tested, in order to observe the interplay of inertio-elastic effects near the intake of the pressure taps. The microdevices were made out of PDMS and the main flow channel, to which the pressure taps were attached, had a cross-section of $270 \times 100\,\mu\text{m}$. Elastic effects were observed in the form of reduced streakline curvature near the taps intake and increased vortex formation.

1 Introduction

On a previous work, we focused on the actual accuracy of different pressure tap configurations for measuring static pressure inside microfluidic devices with a Newtonian fluid, and ended up proposing a novel optimal design [1]. But we did not analyse to what extent the design of the pressure tap would affect the flow. We now turn our attention to this possible issue, in an attempt to characterise the degree of disturbance introduced into a flow of both Newtonian and viscoelastic fluids by attaching a couple of pressure taps to a microfluidic channel.

F. J. Galindo-Rosales et al. (Eds.): IBEREO 2019, SPM, pp. 177–180, 2020.
https://doi.org/10.1007/978-3-030-27701-7_36

2 Experimental

2.1 Channel Geometry and Fabrication

Long straight rectangular (270 × 100 μm) microchannels with a pressure tap on each side were designed and fabricated out of polydimethylsiloxane (PDMS), using standard soft-lithography techniques [1]. Three sets of pressure taps with different intake widths were considered: 54, 108 and 162 μm (w_1, w_2 and w_3, respectively). Figure 1 shows how these are attached to the main channel.

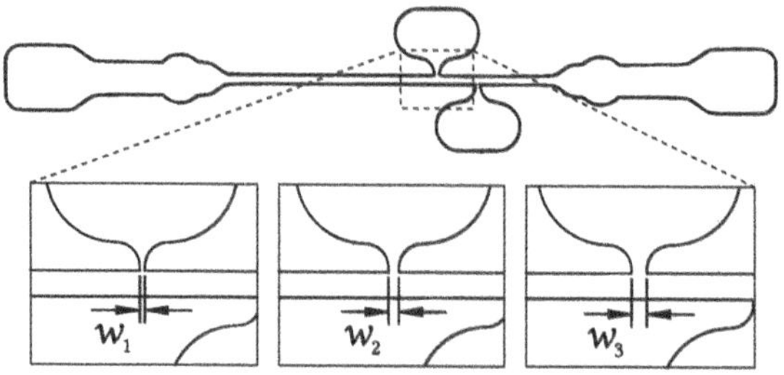

Fig. 1. Schematic representation of the three different test sections

2.2 Fluid Rheology

A Newtonian (D68) and a viscoelastic shear-thinning fluid (DX1000) were tested (see Table 1 for properties). The viscosity curves of the two working fluids were measured at steady shear rates using a stress-controlled shear rheometer (Anton Paar Physica MCR 301) equipped with a plate-plate geometry (50 mm diameter and 100 μm gap). The relaxation time of the viscoelastic solution, λ, was determined by capillary breakup extensional rheometry, using the plate separation drive unit of a HAAKE™ CaBER™ 1 apparatus and a high-speed camera to capture the fluid filament thinning. In order to minimise any inertial effects, the Slow Retraction Method (SRM) [2] was employed: a slow (~0.1 mm s^{-1}) linearly increasing separation of the end-plates is used to initiate the filament breaking process. This way, the elongational flow evolves at a rate orders of magnitude faster than the one at which the plates are moving (parallel plates diameter: 4.0 mm; initial/final gap: 3.0/5.3 mm). All measurements were carried out at ~20 °C.

2.3 Dimensionless Parameters

The Reynolds number characterising the flow in the microchannel is given by: $Re_c = (\rho U_m D_h)/\eta_\infty$, where ρ is the fluid density, η_∞ is the constant viscosity and the plateau shear-viscosity at high shear rates for the Newtonian and viscoelastic fluid, respectively, U_m is the mean velocity in the channel and D_h is

Table 1. Composition and rheology of the test fluids. The concentration of DMSO is given in wt. % in distilled water

Acronym		XG[a]	DMSO[a]	$\rho\,[\mathrm{g\,cm^{-3}}]$	η_∞ [mPa s]	λ [ms]
D68	DMSO	–	68%	1.092	4.0	–
DX1000	XG + DMSO	1000 ppm	52%	1.074	4.5	11 ± 1.3[b]

[a] XG = xanthan gum, DMSO = dimethyl sulfoxide
[b] 95% confidence interval of the sample mean, $\bar{\lambda}$

its hydraulic diameter. The Reynolds numbers considered ranged from ~0.05 to ~5 for both working fluids.

For the viscoelastic fluid only, the characteristic Weissenberg number is defined by: $Wi_c = \lambda\dot{\gamma}$, where λ is its extensional relaxation time and $\dot{\gamma}$ the shear rate, which can be expressed as $\dot{\gamma} \sim Q/\left(wd^2\right)$ for a rectilinear channel of width w and depth d ($^w/_d \gg 1$) [3]. The deformation rate ($\dot{\gamma}$) ranged from $\sim 17.2\,\mathrm{s^{-1}}$ to $\sim 1905\,\mathrm{s^{-1}}$ and the Weissenberg numbers were $\sim 0.19 \leq Wi_c \leq \sim 21$.

2.4 Flow Visualisation

The working fluids were seeded with 2 μm fluorescent tracer particles (Molecular Probes® FluoSpheres®, F8825), at a concentration of ~1000 ppm. 0.1% $^w/_w$ of

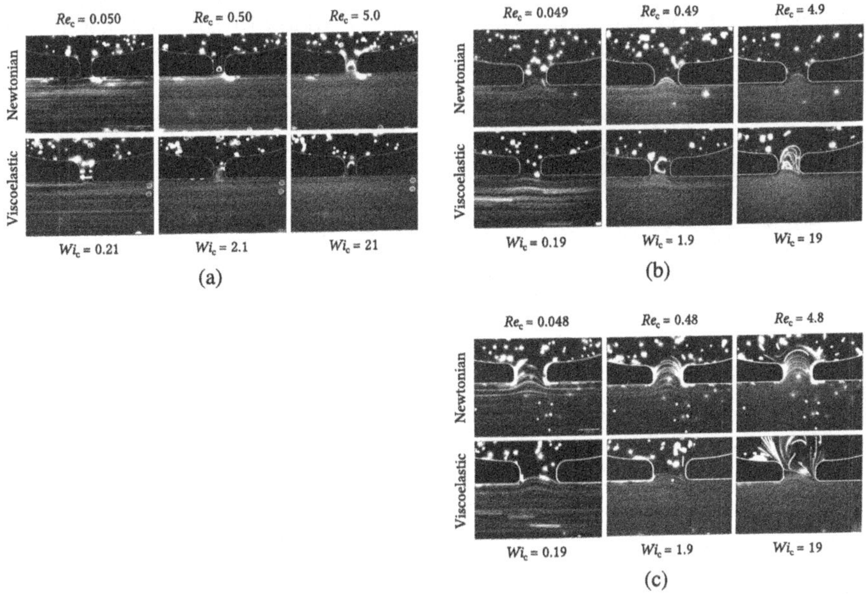

Fig. 2. Streakline images of flow passing by a pressure tap of intake width: (a) $w_1 = 54\,\mu\mathrm{m}$ (b) $w_2 = 108\,\mu\mathrm{m}$ (c) $w_3 = 162\,\mu\mathrm{m}$

sodium dodecyl sulfate (SDS) (Sigma-Aldrich) was added in order to minimise the adhesion of the particles to the channel walls.

The optical setup consisted of an inverted microscope (Leica Microsystems GmbH, DMI 5000 M) equipped with a sensitive monochromatic CCD camera (Leica Microsystems GmbH, DFC350 FX), an external light source for fluorescence excitation (100 W mercury lamp) and a filter cube (Leica Microsystems GmbH, A: excitation filter BP 340–380; dichromatic mirror 400; suppression filter LP 425). All streakline images were taken at the midplane of the microchannels. Figure 2 shows the flow patterns in the three geometries studied for selected Reynolds/Weissenberg numbers.

3 Conclusions

With the Newtonian fluid the flow patterns present a much bigger curvature in the region of the pressure tap intake, reaching far into its interior for the case of $w_3 = 162\,\mu\text{m}$. With the viscoelastic fluid one observes larger vortices, especially at the higher Reynolds numbers, but less curvature near the pressure taps, since the elasticity of these fluids dampens such inertial effects. The pressure tap of Fig. 2a revealed greater propensity for vortex formation, particularly in the case of the Newtonian solution. Moreover, the narrower the intake, the less the flow in the main channel was visibly affected.

Acknowledgements. This research was funded by FEDER (COMPETE 2020) and FCT/MCTES (PIDDAC), grant number POCI-01-0145-FEDER-030764. J.H.G.-O. also acknowledges "Programa de Fomento e impulso de la actividad investigadora de la Universidad de Cádiz 2018/19".

References

1. Rodrigues, T., Galindo-Rosales, F.J., Campo-Deaño, L.: Materials **12**, 1086 (2019)
2. Campo-Deaño, L., Clasen, C.: J. Non-Newton. Fluid Mech. **165**, 1688 (2010)
3. Pipe, C.J., McKinley, G.H.: Mech. Res. Commun. **36**, 110 (2009)

Author Index

F. J. Galindo-Rosales et al. (Eds.): IBEREO 2019, SPM, pp. 181–182, 2020.
https://doi.org/10.1007/978-3-030-27701-7

The manufacturer's authorised representative in the EU is Springer Nature Customer Service Centre GmbH, Europaplatz 3, 69115 Heidelberg, Germany. If you have any concerns regarding our products, please contact ProductSafety@springernature.com

Printed and bound by CPI Group (UK) Ltd, Croydon, CR0 4YY
30/09/2026
02225495-0004